My Pet
CHICKEN
Handbook

My Pet CHICKEN *Handbook*

SENSIBLE ADVICE AND SAVVY ANSWERS FOR RAISING BACKYARD CHICKENS

LISSA LUCAS & TRACI TORRES
THE EXPERTS AT MY PET CHICKEN

RODALE

Rodale books may be purchased for business or promotional use or for special sales.
For information, please write to:
Special Markets Department, Rodale Inc., 733 Third Avenue, New York, NY 10017

Printed in the United States of America

Rodale Inc. makes every effort to use acid-free ∞, recycled paper ♻.

Photographs by Cristian Balty/E+, page 182; Michael Blann/Digital Vision, page 154; Daily Herald Archive/SSPL/
Getty Images, page 3; Rachel Fiore, page 81; Nick Gould (www.nickgould.net), page 17 (#15); GK Hart/Vikki Hart/
Stockbyte, page i; Zena Holloway, page 5; Jean-Philippe Ksiazek, page 8; Lissa Lucas, pages 38, 52, 57, 63, 89, 100,
103, 108, 123, 134, 159, 185; Derek Sasaki, pages ii, vi–vii, 16–17 (except #15), 31, 34, 41, 45, 47, 51, 68, 70, 72, 86,
91, 97, 98, 106, 110, 112, 113, 115, 121, 128, 136, 151, 167, 174, 179; Johanna Siegmann, page x; Paul E. Tessler/Photodisc,
page 93; Christopher Testani, background on cover and on pages iii, 1, 53, 101, 155, 183; Traci Torres, pages 12, 29, 32,
118, 163, 186; Kathryn Weller, pages 232–233.
Book design by Chris Rhoads
On the cover: Silver Laced Wyandotte bantam hen

Library of Congress Cataloging-in-Publication Data is on file with the publisher.
ISBN 978-1-62336-001-6 paperback

Distributed to the trade by Macmillan
2 4 6 8 10 9 7 5 3 1 paperback

RODALE.

We inspire and enable people to improve their lives and the world around them.
rodalebooks.com

This book is dedicated to my pet chickens,
including Lily and Galatea, Wheaten Ameraucanas
who taught me how affectionate and smart chickens can be;
Gautier, a Salmon Faverolles rooster who taught me
that fierce and gentle are not mutually exclusive qualities;
and Hildy, a blind Speckled Sussex hen
who taught me never to give up.

—LISSA LUCAS

To my family:
you are my center of gravity;
and to my chickens:
I couldn't imagine life without you.

—TRACI TORRES

Contents

INTRODUCTION viii

PART 1: **Planning for Chickens Like an Expert**   1

CHAPTER 1: **A BRIEF HISTORY OF CHICKENS**   2

CHAPTER 2: **THE BACKYARD CHICKEN BOOM**   9

CHAPTER 3: **CHOOSING BACKYARD BREEDS**   14

CHAPTER 4: **TRUTH IN (CHICKEN) ADVERTISING**   28

PART 2: **Practical Housing Arrangements**   53

CHAPTER 5: **FLOCK-MANAGEMENT STYLES**   54

CHAPTER 6: **THE BROODER AND COOP**   66

CHAPTER 7: **PREDATORS AND PESTS**   80

PART 3: Starting and Caring for Your Flock   **101**

CHAPTER 8: **ESTABLISHING YOUR FLOCK**   **102**

CHAPTER 9: **NURSERY TO COOP**   **117**

CHAPTER 10: **CARE OF ADULT CHICKENS**   **138**

PART 4: Determining When There Is a Problem   **155**

CHAPTER 11: **RECOGNIZING SIGNS OF ILLNESS**   **156**

CHAPTER 12: **BEHAVIOR AND MISBEHAVIOR**   **176**

PART 5: Reward Time!   **183**

CHAPTER 13: **ENJOYING YOUR HOME-PRODUCED EGGS**   **184**

CHAPTER 14: **RECIPES FOR EGG LOVERS**   **189**

CHAPTER 15: **TREATS FOR YOUR CHICKENS**   **227**

ACKNOWLEDGMENTS 232   ABOUT THE AUTHORS AND ABOUT MY PET CHICKEN 234   INDEX 235

Introduction

In 2004, while planning for my very first flock of chickens, I began to form the idea for My Pet Chicken. There was so little how-to information on the web for novice chicken keepers. There wasn't a single e-commerce web site dedicated to hobbyists; how would people figure out which chicken breeds to get, among hundreds of choices? How could they painlessly determine whether they'd have the time, energy, and money to care for their own flock? The utter lack of information on the web inspired me.

But the idea was a long time incubating. While I wasn't sure if it might just be another wacky idea with little chance of succeeding (a pet chicken business, after all!), my husband Derek increasingly came to believe in it. Despite many friends thinking we were, well, *clucking crazy*, his persistence finally won me over.

Today, My Pet Chicken is a multimillion-dollar business, and our web site receives millions of visitors annually. The easy access to top-notch information, photography, and products on our web site helped to *create* the backyard chicken movement that has now swept the nation. We are so very proud to be a part of that and to be connecting people with chickens every day.

So why a book if our web site has done such a thorough job of introducing people to life with chickens? Well, after 8 years in business and after having hand-held 70,000 customers through their chicken-keeping experiences, we've learned a few things. This is the ultimate insider's guide to keeping chickens, and we explain everything you need to know to get started—all conveyed to you with our unique insight into the mind of a beginning chicken keeper. We also teach the incredibly important stuff nobody else is talking about. Stuff like what it means to "cast" your flock and how to do it like a pro; rookie mistakes to avoid when ordering chicks; how to think like a chicken so you can understand their behavior and culture; how to encourage good "cold weather chickens," and more. Countless more chicken secrets. And we're happy to share.

Starting with chickens is thrilling, and I wish you success and enjoyment for years to come. I'm confident that, armed with what you'll learn here, you'll be beautifully prepared for your chicken adventure. And of course, should you have any questions, I welcome you to e-mail us at info@mypetchicken.com. A real, live, chicken-loving expert will write you a personalized response, usually within a day.

—Traci Torres

Are You Really Ready for Chickens?

If you have all of these questions covered, that's great—you're well prepared! If not, don't worry. This book will guide you through all of the major decision points, and you can revisit this checklist when you've finished the book.

TOWN AND NEIGHBORS

✓ Do your town's zoning regulations allow for chickens?

✓ Does your town's department of health have regulations regarding chickens?

✓ Are you prepared to address any concerns neighbors may have?

CHICKEN COOP AND RUN

✓ Have you considered whether to purchase a chicken coop, build new, or adapt an existing structure (like a shed) and how these choices will impact your budget?

✓ Have you determined the best site for your coop, taking into account sun, shade, grading, and proximity to electricity and water?

✓ Have you thought about how a coop and run may affect your yard's available space, foot-traffic flow, and its overall appeal?

✓ If you're planning to purchase a coop, have you considered that it may take several weeks to build and ship?

✓ If you're planning to build a coop or adapt an existing structure, have you considered the design features that will best suit your needs and your flock's needs?

✓ If you have dogs, have you considered the necessary choices you'll face in case the dogs can't be socialized with your chickens?

✓ Have you educated yourself about predators in the area and how you'll protect your flock against them?

✓ Are you prepared to purchase metal containers to store chicken feed and have you considered where you'll store feed so it doesn't attract rodents?

FLOCK CONSIDERATIONS

✓ Eggs, chicks, juveniles, or rescue chickens—have you decided how you want to start your flock?

✓ Have you researched which chicken breeds or mix of breeds will match your goals, your locale, and your setup?

✓ Do you have a backup plan in case of unwanted roosters?

✓ Is there an avian vet or farm vet in your area who can treat your chickens in case of illness or injury?

✓ Is there a local feed store where you can purchase supplies, or will you need to source them online?

✓ Have you considered which flock-management style suits you?

ONGOING CARE AND COMMITMENT

✓ Will you make time to collect eggs, make sure waterers and feeders are full, and check on your chickens each day?

✓ Can you commit to cleaning or refreshing your flock's bedding once per month?

✓ Is someone in your family willing to do a thorough sanitizing of the (messy) coop once or twice per year?

✓ Is there someone you can rely on to take care of your flock when you are out of town?

✓ Will someone be available to take your chickens to the vet in case of emergency?

Planning for Chickens Like an Expert

Backyard chickens aren't new. A century ago, the idea of keeping chickens in your backyard wouldn't have caused anyone to raise an eyebrow or bat a lash. Chickens have been domesticated for thousands of years. Of course, we could have a long, intellectual discussion about where and when they were first domesticated, but let it suffice to say that chickens have been around for a *long* time, even in America.

In this section of the book, we'll look at the history of chicken keeping in America and delve into the qualities you'll want to consider when choosing the best chicken breeds for your home flock.

A BRIEF HISTORY of CHICKENS

It probably comes as no surprise to hear that chickens
came here shortly after the arrival of the first English settlers
at Jamestown. And much has changed with chickens and
the business of chickens over a few hundred years.

Even then, chickens were more of a sideline than a business or vocation. They weren't exactly regarded as farm animals—not in the same way cattle or horses were. Unlike those animals, chickens took up little space and required minimal care. Chickens were kept in small flocks, so it was usually very much a family affair. For instance, it was traditionally the job of the youngest child to gather the daily eggs, and, for the most part, the family chickens were managed by mom rather than dad. Extra eggs were sold at the market for "egg money," a little supplemental income for mom, in addition to the fresh eggs provided for the family.

In America, chickens were historically more of a hobby—and they were a *very* American hobby,

at that. For example, Thomas Jefferson gifted his favorite granddaughter with a pair of bantams for her pleasure. Teddy Roosevelt owned a one-legged rooster, and his children had pet hens with names like Fierce and Baron Spreckle. It's only been recently that chickens have gotten a bad reputation. Among some people, they are regarded as dumb or dirty birds. However, nothing could be further from the truth, though it's not hard to figure out where these erroneous ideas came from if you look at the conditions in factory farms today. Certainly, if it were permissible to keep dogs or cats (or any animal) in similar conditions, our ideas of their intelligence and cleanliness would be different as well.

But the reputation of chickens was once quite different—in the past, they were symbols of goodness and bravery. In medieval times, for instance, the call of a rooster was thought to drive away demons and devils. Chickens have been variously considered symbols of pride, fertility, and even enlightenment, being heralds of the dawn. Roosters faithfully announced the break of day, so sunlight would illuminate the world and send evil and darkness slipping off into the shadows. A rooster was the model of an excellent father and, in fact, a symbol of fatherhood: He was vigilant in protection of his family, genteel in demeanor, and a generous provider. After all, when a rooster finds a choice morsel of food, he calls his family to the feast, letting the hens eat first while he watches for danger and keeps the peace. Roosters are even graceful dancers; consider the elegant sweep of a rooster's extended wing as he moves in circles around a hen he's trying to charm.

It wasn't only roosters that were esteemed, though. Hens were also admired—symbolic of courage, self-sacrifice, and domestic virtue. Jesus compared himself gathering the tribes of Israel to a mother hen gathering chicks under her wings. Hens sit faithfully for weeks on the nest to hatch their eggs, barely eating or drinking until their little chicks have emerged. Then, there are 5 to 6 weeks more of care until the young birds are ready to be on their own. For good reason, hens were regarded as faithful, protective mothers—the epitome of tenderness and motherly love. In fact, the Greek historian Plutarch described the hen in this way: "What of the hens whom we observe each day at home, with what care and assiduity they

A 1933 snapshot shows a group of young women with their chickens.

govern and guard their chicks? Some let down their wings for the chicks to come under; others arch their backs for them to climb upon. There is no part of their bodies with which they do not wish to cherish their chicks if they can, nor do they do this without a joy and alacrity which they seem to exhibit by the sound of their voices."

Even chicken eggs were symbols of rebirth and the cycle of life. In many cultures they were decorated in the springtime, joyously sought and found, a herald also of the returning plenty that came with the change of seasons. People have been delighting in chickens for thousands of years.

For early American settlers, chickens were a great family hobby. You didn't have to live on a farm to keep chickens. Cows, horses, goats, and other livestock required large amounts of open pasture and a lot of work and expense. By contrast, chickens could live in a small yard. They cost little, and they required little effort. They would eat grains dropped by other animals or scraps from the kitchen, so they reduced waste. They even cut down on the insect population in their immediate vicinity with their sharp-eyed bug hunting. They were self-sufficient to a large extent, and often—although we don't recommend this today—the family chickens weren't even provided with a coop or other housing (like a barn). They simply roosted in trees or wherever they could, left to avoid predators on their own—or not.

That's how it was for chickens in the early years of the United States.

THE CHICKEN CRAZE BEGINS

In the early to mid-1800s, for the first time, a breed of chickens known as Cochins was imported from China, and these appealing, fat, fluffy balls of feathers created a sensation. Compared to the flighty, quick birds most families kept for eggs, Cochins seemed as docile as kittens, and they were beautiful, too, with elaborately feathered legs and silver-laced, blue, and mottled colors. They made exceptional mothers that were also calm and easy to handle, so the chicks they raised were often especially docile, too.

In short order, everyone wanted Cochins. But Cochins couldn't really be kept in the same catch-as-catch-can way that other chickens were being kept at the time. For one thing, their ponderous size made it improbable, if not impossible, for them to be able to roost high up in trees and nimbly avoid predators on their own. For another, their profusely feathered legs could be problematic in muddy environments. And lastly, they were expensive in comparison to other chickens, so there was more of a desire to provide housing and security for the rare, new China Cochins.

As the first real exhibition chickens in the United States, Cochins set off a craze for poultry exhibition that has continued to this day. Professional farmers began getting involved with chickens, using selective breeding to create chickens with fancy feathering: crests, beards, feathered legs, and unusual plumage. However, chickens bred in this way didn't necessarily breed true;

offspring might be completely different in character from their parents. With no breeding standards, there could be no real breeds like there are today. Establishing a new breed is a complex process. Once you have produced a stable cross, other breeders must show interest and standards (what the bird should look like) must be agreed on. In the early 19th century, those standards did not exist yet.

THE MODERN CHICKEN

The advent of modern poultry keeping resulted from a few different things happening at once, most notably the creation of the American Poultry Association (APA) in 1873. The APA began standardizing breeds, so that to be called a "breed," a chicken would have to be able to produce the same type of bird in future generations. This meant that someone buying chickens could be reasonably assured of the type of chickens he'd see in future generations, and could choose breeds based on qualities needed, such as cold hardiness, productivity, or early maturity. This was also the year that the first artificial incubator was patented—so chicks could now be hatched in quantity.

Once there were standard breeds as well as a way to produce chickens in quantity, the chicken world was transformed. More people had access to keeping chickens, since hatching chicks no longer relied so much on Mother Nature. You didn't need broody hens to hatch eggs (broodiness could even be bred out, if desired). These changes also made it possible for chicken and egg businesses to grow substantially. Today, a single commercial incubator can hatch many thousands of chicks—a vast difference from the measly dozen or so eggs at a time a broody hen can hatch.

Other changes followed soon after: Mass transportation allowed baby chicks to be shipped by train as early as the 1890s, and hatcheries began springing up all over the country. Farmers could order chickens "designed" to mature quickly and lay lots of eggs, and hobbyists could branch out into rare and fancy breeds they'd been coveting for years simply because there was better access to them. But while improved transportation helped the hobby, it also enabled chicken keeping to take its first steps toward becoming a large-scale agribusiness.

Egg businesses were now able to order hundreds and thousands of baby chicks. And because of this, the birds began to be kept in different conditions. After all, when you have a flock of hundreds or thousands rather than just 20 or 30, your chickens will need much more space than what had been dedicated to small hobby flocks to allow them to forage enough to sustain themselves. And since the leftover table scraps of a family of four (or six or 10) wouldn't do much to supplement the foraging of that many birds, commercial chicken feed was developed, and it was formulated to keep chickens well nourished and maximize the number of eggs they could lay. Around the turn of the 20th century, chickens laid only about 30 eggs per year. Now, good layers are expected to produce around 200 to 250 eggs per year—seven to eight times as many. The selective breeding and the improved nutrition are paying off.

The Downside of Modern Methods

To remain competitive, egg farms had to become very efficient, so they changed their management practices: Confinement became the norm as producers began to realize savings from it. Confining chickens required less space, and it meant that they were in less danger from predators and eggs could be more easily gathered. Eggs were so cheap at the grocery store that fewer people kept their own chickens for eggs, and the space factory farms allocated to their birds grew smaller and smaller.

Since chickens don't naturally live in close quarters, confinement created new problems, from aggression to increased risk of disease. Commercial chicken feed had started out by providing much-needed balanced nutrition for flocks that were confined away from natural forage; however, the focus of factory farming operations came to be on the quantity of eggs produced, rather than on the quality.

Stress made the huge flocks vulnerable to many illnesses, and the density of their living spaces made it easy for sick birds to pass those illnesses to other birds. Bored birds would pick on each other ruthlessly from the stress of living in cages or cramped quarters their entire lives. Birds at the bottom of the pecking order couldn't get away from more aggressive hens like they could if allowed to roam free, and aggression became a serious problem. Rather than make the determination that chickens need a certain minimum amount of space for their physical and mental well-being, research went in the other direction: How can we squeeze more chickens into smaller spaces? But overcrowding causes stress, which in turn causes aggression among chickens.

Today, the most common "solution" for the stress and aggression caused by close confinement in large factory farms is to debeak young chicks, a painful process often referred to with the less horrifying term *beak trimming*.

As someone interested in backyard chicken keeping, you probably don't need to hear from us about the awful conditions chickens endure in many big commercial farming operations. You'll want to raise your chickens in humane, healthy conditions. And you will have the chance to discover how smart and endearing they really are. Fortunately, some progress, however modest, has been made in chicken farming. The European Union Council Directive has banned wire battery cages in the European Union. California has taken some steps as well, due to consumer outrage. However, some other states are taking steps backward, and animal abuse continues in

Egg-cetera!

In most breeds, juvenile roosters can be distinguished from juvenile hens by the shape of their hackle and saddle feathers. Hens have rounded feathers, while roosters have long, pointy feathers in comparison.

factory farms. Your chickens will have a much better life.

Restoring Common Sense

The *best* solution to the aggression caused by close confinement is pretty simple: Give chickens more space and allow them to behave like chickens, rather than treating them like egg-making machines. When you keep a backyard flock, you have control over how the birds are kept and what they're eating. And by raising chickens that produce eggs in your own back-yard, you're also reducing demand for inhumanely produced factory farm eggs, not to mention that you have more control over what you are feeding your family.

Not coincidentally, giving chickens room and access to pasture increases the healthfulness of their eggs, on top of all the other benefits. *Mother Earth News* studies in 2007 and 2010 concluded that eggs from chickens raised on pasture are higher in omega-3s; vitamins A, D, and E; and beta-carotene while being lower in cholesterol and saturated fat. Really, this is what the new backyard chicken movement is all about.

The BACKYARD CHICKEN BOOM

Keeping chickens-even small flocks in a family setting-is nothing new,
as you have just learned. Even though backyard chickens are
now considered a novel trend-news that has inspired many eggzasperating
puns from newscasters and journalists-in reality, this "new" trend
greatly resembles the old way of raising chickens.

Chickens are again being kept in small flocks by families, with Mom and Dad and the kids all pitching in together. What differs today is that families are now taking advantage of the same improvements that factory farms took to such an extreme. For instance, we now have the technology to transport small numbers of baby chicks safely to your home for a "micro-flock" of just three or four. And families provide these little flocks with secure housing and nutritionally balanced feed. The chickens don't have to fend for themselves.

Because they are getting such good care, just a few birds can deliver enough eggs for a small fam-

ily year-round. Most of us wouldn't be happy getting only 30 eggs a year per hen—and only in the spring! But we now know so much more about how to provide good care. Those of us who keep pet chickens are in the best position to take advantage of all this information: Today's small-scale chicken keepers provide good nutrition, secure shelter, and plenty of space so that they'll have a happy, healthy flock over the long term—a flock that produces eggs that are far more nutritious than what you can get from any commercial operation.

As a result of these improvements—and also as a reaction to factory egg farming—backyard chickens are certainly becoming more popular.

WHY BACKYARD CHICKENS?

Here are the top 10 reasons more people are turning to backyard chickens.

Chickens make great pets. They have personality galore, and they're extraordinarily easy to care for. They're bright, funny, quirky, friendly, loving little balls of feathers—and they're entertaining, too. When you have a flock, you'll find they have their own friends, their own cliques, their own favorite nests. Chickens come in such an array of colors, patterns, shapes, and sizes that some of them look more like exotic tropical birds—or even alien life-forms—than farm animals.

Keeping chickens is a lifestyle choice; you keep chickens if you want to try to live in a more sustainable way. Having chickens helps fulfill a positive, back-to-the-farm way of living that's about becoming more sustainable. It's also a way to celebrate local, slow food, and reestablish a constructive connection with your neighbors and your neighborhood.

It allows you to have more control over the type of food you put on your table. You want organic? You want non-GMO (genetically modified organism)? You want cruelty-free? These choices are all yours to make when raising your chickens.

Chickens will eat your table scraps and convert them into eggs on the one hand, and fertilizer on the other. If you grow vegetables or flowers, you'll find that composted chicken manure is great for your home garden, adding organic matter and nutrients to the soil. Plus, chicken manure from layers tends to be relatively high in calcium, which is helpful for plants, warding off blossom-end rot on tomatoes, for example.

What Made You Want to Keep Chickens?

In May 2010, My Pet Chicken polled its e-mail newsletter subscribers to discover which factors mattered most to would-be chicken keepers.

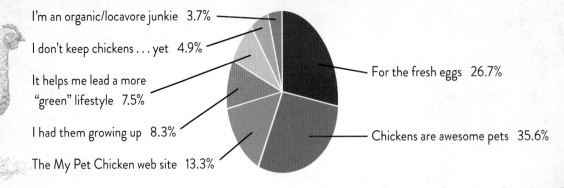

I'm an organic/locavore junkie 3.7%

I don't keep chickens . . . yet 4.9%

It helps me lead a more "green" lifestyle 7.5%

I had them growing up 8.3%

The My Pet Chicken web site 13.3%

For the fresh eggs 26.7%

Chickens are awesome pets 35.6%

Chickens will cut down on the number of insects in your yard. Anywhere chickens are allowed to forage, they'll snap up spiders, ticks, beetles, grubs, worms, grasshoppers, and more. They love to dig through lawn clippings and yard waste, too.

The eggs from hens raised with access to your backyard will be tastier and more nutritious! Research shows they're not only higher in omega-3s, beta-carotene, and vitamins A, D, and E, but they're lower in cholesterol and saturated fat.

They taste better, too. It's something you can see: All that extra nutrition gives backyard eggs a dark orange yolk—not the pale yellow color you see in store-bought eggs.

You'll be eating really fresh eggs—sometimes just minutes old—as opposed to the eggs you get in a grocery store, which can be 6 weeks old or more.

You'll be giving your children positive values. Just as with other pets, keeping chickens can help kids learn about responsibility. But because chickens give back in such a tangible way—eggs!—your kids can also learn about reciprocity and how the care they provide impacts their pets directly. Once they taste the eggs, they'll also come to learn that store-bought isn't always better. Some things are worth doing yourself.

You'll have control over how humanely your wonderful egg producers are treated— and how healthy and clean their environment is.

Chickens are so easy to care for. No walking, no pooper-scoopers, no grooming, no boarding when you go away; they won't scratch up your furniture or chew your favorite slippers.

PREVENT PROBLEMS BY MAKING SMART CHOICES

You're interested in backyard chickens or you wouldn't be reading this book. My Pet Chicken has been in the backyard chicken business for some years, and we've seen how easy it is for newbies to make mistakes early on that will lead to problems later. We hear from people every day who have read other backyard chicken how-to books, but are still dealing with easily avoidable concerns and frustrations. Let's address a few of those issues up front, so you can start your new hobby in the best possible way.

One of the worst things you can do is to jump into buying chickens before thinking the whole situation through. Of course, this goes for any pet—clearly, you don't want to acquire a pet before you're fully prepared. But it's especially important to consider when you're talking

Egg-cetera!

Blue eggshell color comes from biliverdin; brown eggshell color comes from protoporphyrins. Pink eggs are the result of small quantities of a rosy brown pigment, while green eggs have both blue and brown pigments.

about pet chickens. For starters, supplies may not be as easy to get as they are for other pets. Depending on where you're located, it's not going to be just a matter of running to the grocery store to pick up food for your chickens like you can for your cats. And sleeping arrangements for your flock won't be as simple as getting a doggie bed at the local Target. Plus, you may not have chicken-keeping friends or family that you can turn to for advice if you run into any problems. Further, in many areas, veterinarians who treat chickens can be hard to find.

But making a reservation for your baby chicks is easier—and more tempting—than buying a dog or cat. Purebred chicks are inexpensive, just a few dollars each. Compare that to the cost of purebred dogs or cats, which can easily cost thousands. With puppies or kittens, you'll have to find a breeder relatively close by, or make special (expensive) arrangements for transport. Even adopting dogs or cats at the local animal shelter usually requires an adoption fee and an agreement to have your pet spayed or neutered. By contrast, you can order baby chicks online for as little as $2 to $3 per chick, so impulse buying is more likely to happen. Or you can stop by your local farm and feed store and buy chicks on the spur of the moment. Some people wrongly think it's a great idea to give chicks as a gift for Easter, and will rush out to buy them without considering all their needs.

It's not that chickens are hard to care for. They're not. At all. In fact, they're easier to keep than most other pets. You don't have to take chickens for daily walks or let them out to poo.

Free-ranging your flock is wonderful, but check with your town's zoning board first to be sure it's permitted.

They don't get separation anxiety. You don't have daily litter-box cleanings or twice-a-day feedings, and you don't have to worry about claws being sharpened on the sofa. There are no brushings or groomings. No baths. No indoor shedding. Even purebred chickens are inexpensive to buy, feed, and keep. Plus, they make a great alternative for people who are allergic to cats and dogs.

However, chickens—particularly baby chicks—do require special care. For instance, chicks need to be kept in a heated brooder until they're old enough to transition to the big-girl coop. (This is also one reason why putting chicks in an Easter basket for the kids is not a good idea—at room temperature, they'll die of hypothermia without the warmth they need.) So you can't, in good conscience, simply bring chicks home on an impulse and *then* try to figure out how to care for them. You need to prevent these problems by making smart choices, so you're armed with knowledge before you even begin. (And, of course, that's why you're reading this book.) Consider the following:

First, determine if you can make the time commitment necessary for proper care. Chickens are probably among the easiest pets to care for, but you'll still need to commit to providing them with a secure area, proper food, clean water, shelter from weather extremes, and other necessities. Chickens need to be checked on daily, and there are monthly and semiannual chores, too. We discuss the details of proper chick and chicken care starting on page 117.

Next, determine if you can provide proper housing. Baby chicks need a brooder; adult chickens need a coop. A brooder is usually located in your home somewhere, while your coop will be located outside. City dwellers may keep chickens on rooftops or patios, while those in suburban areas will probably locate coops in their yards. We discuss the details of arranging your brooder and coop in Chapter 6.

Last but not least, make sure chickens are permitted in your area. It is illegal to keep chickens in some places. So before you do anything else, you'll want to contact both your local health and zoning boards to see if there are any regulations relating to keeping chickens in your area. If there is a homeowner's association for your neighborhood, you'll also want to contact them to find out if they have additional regulations. Sometimes one group will have regulations even when the others don't—that's why you want to contact all the local authorities that can regulate what you do with your property.

Be thorough about it, too.

If you're told that chickens aren't allowed, ask to read the specific statutes. The reason is that sometimes chickens are assumed to be prohibited when regulations actually do permit them. What is often in question is your area's definition of livestock. Some towns may outlaw livestock within city limits—but chickens often meet the definition of a pet (like a potbellied pig might).

Again, make sure you check all this out before you purchase any equipment, a coop, or other supplies. You don't want to spend a bunch of money having a coop shipped to you—or waste money and time building your own coop—if you aren't even allowed to have chickens. It's *especially* important to be sure of local regulations before making any reservations for chicks. Keep in mind: If you've placed an order for chicks and then have to cancel it due to local regulations, your hatchery will need time to find new adopters for the chicks you reserved and then discovered you couldn't care for after all.

CHOOSING BACKYARD BREEDS

When it comes time to choose what breeds will
comprise your new backyard flock, it's easy to get overwhelmed.
With adequate background information, though,
we're sure you'll find the perfect match for your situation.

There are numerous chicken breeds and varieties, many of which are great for a backyard situation. There is no *one* perfect breed for everyone; after all, your needs will vary.

Some people will want birds that are great layers, while others will want those that make for excellent pets (with eggs as a wonderful bonus). And some will want birds that are both. Some will look for birds that are good foragers, or that have some natural camouflage, and would thrive in free-range situations. Others may want birds that do well confined in a small suburban yard. Some may live in the extreme North and will need birds that are especially cold hardy, while others may live in the desert areas of the Southwest, where birds need to be able to weather sweltering conditions. If you're set on exhibiting your chickens, you'll want to stay away from crossbreeds and birds that can't be shown. Which chicken breed is best for you is dependent on many things, not the least of which will be your family's own preferences and needs. That's not to say that there aren't breeds that will work for all of these different goals—there are, but there are many that won't, so you want to make the right breed choices up front to avoid disappointment later.

To help you sort out and compare the characteristics of different types of backyard chickens, we've created a helpful Breed Selector Guide.

Take a few moments to glance over the guide

and get some initial impressions of the many breeds available. When choosing the breeds that will make up your new flock, you want to pick ones that you like, naturally! But you also want to think of yourself as a sort of casting director. You have a certain amount of control over the personalities in your flock, so try to cast the best possible breed or mix of breeds for your family.

Making a choice can be difficult, though, especially when you're new to chickens. After all, there's a lot of jargon you may not even be familiar with. "Broody"—what's that? Is it something you want, or is it something to avoid? If you get cold-hardy birds, does that mean you'll have plenty of winter eggs? What does "tolerant of confinement" really mean? You can learn more about the various characteristics highlighted in the guide starting on page 18.

The Wild World of Chickens!

Once you've determined the breed characteristics that will be best for your needs (which we'll walk you through in the pages ahead), you can start designing the flock you've been dreaming of. Do you love birds with floofy, profuse feathering? Or do you prefer tight, hard feathers and a steely gaze? Perhaps you only want blue and lavender birds . . . or perhaps you're like most beginners, and you'd like to see a veritable Noah's ark of chicken shapes, colors, and sizes strutting around your yard. Whatever your proclivities, we can guarantee you'll have a blast selecting your first flock. How could you not, with the roster of several hundred breeds and varieties to choose from? The gallery that follows is a tiny sampling of what the chicken world has to offer.

1. White Sultan hen
2. Golden Sebright Bantam rooster
3. Buff Orpington rooster
4. Self Blue Araucana Bantam pullet
5. Light Brahma Bantam pullet
6. Tolbunt Frizzle rooster
7. Wheaten Asil hen
8. White Showgirl Silkie Bantam cockerel
9. Appenzeller Spitzhauben hen
10. Easter Egger hen
11. Mille Fleur d'Anvers Bantam hen
12. Black Langshan rooster
13. White Frizzle Polish hen
14. Self Blue Bearded Belgian Bantam hen
15. White Yokohama rooster
16. Red Pyle Modern Game cockerel
17. White Faced Black Spanish rooster
18. Serama Bantam rooster

BREED SELECTOR GUIDE

BREED	LAY RATE	EGG SIZE AND COLOR	COLD HARDY	WINTER LAYER
Ameraucana	Good	Medium blue	Yes	No
Ancona	Excellent	Large white	Yes	No
Andalusian	Good	Large white	No	No
Antwerp Belgian	Fair	Extra small white	No	No
Appenzeller Spitzhauben	Good	Medium white	Yes	No
Araucana	Fair	Medium blue	Yes	No
Asil	Poor	Medium tinted	Yes	No
Australorp	Excellent	Large brown	Yes	Yes
Barnevelder	Good	Large dark brown	No	No
Bearded d'Uccle Bantam	Fair	Extra small white	No	No
Booted Bantam	Fair	Extra small white	No	No
Brahma	Good	Large brown	Yes	Yes
Buckeye	Good	Medium brown	Yes	Yes
Campine	Good	Medium white	No	No
Catalana	Very good	Medium white	No	No

Lay rate: Excellent (5 or 6 per week), very good (4 or 5 per week), good (3 or 4 per week), fair (2 or 3 per week), poor (1 or 2 per week)
Egg size: Extra small (1.25 oz or less), small (1.5 oz), medium (1.75 oz), large (2 oz), extra large (2.25 oz), jumbo (2.5 oz or bigger)
Cold hardy: Tolerates cold winter temperatures
Winter layer: Continues laying in winter

HEAT HARDY	FORAGER	BROODY	DOCILITY	CAUTIONS
Yes	Better	Occasional	4	Flyers; slow to mature
Yes	Best	Seldom	1	Flyers
Yes	Best	Seldom	2	Wild; can be noisy
No	Average	Frequent	4	
Yes	Best	Occasional	2	
No	Average	Frequent	3	
Yes	Average	Occasional	2 (difficult for beginners)	Aggressive with other chickens; docile with humans
No	Average	Occasional	3	
No	Better	Occasional	3	
No	Average	Frequent	5	
No	Average	Frequent	4	
Yes	Better	Occasional	4	
No	Better	Occasional	4	Can get picked on in mixed flocks; good mousers
Yes	Better	Seldom	2	Flyers
Yes	Better	Seldom	1	

(continued)

Heat hardy: Tolerates summer heat especially well
Forager: Able to hunt for food
Broody: Wants to hatch eggs
Docility: 5 (potential lap bird), 4 (friendly), 3 (average), 2 (flighty), 1 (avoids human contact)

BREED SELECTOR GUIDE *(CONT.)*

BREED	LAY RATE	EGG SIZE AND COLOR	COLD HARDY	WINTER LAYER
Chantecler	Very good	Large brown	Yes	Yes
Cochin	Fair	Medium brown	Yes	No
Cornish	Poor	Small light brown	No	No
Crevecoeur	Fair	Medium white	No	No
Cubalaya	Very good	Medium white	No	No
Delaware	Very good	Light brown	Yes	Yes
Dominique	Good	Light brown	Yes	Yes
Dorking	Good	Medium tinted	Yes	No
Dutch	Fair	Extra small tinted	No	No
Easter Egger	Very good	Large blue, green, any	Yes	No
Empordanesa	Good	Medium chocolate	No	No
Favaucana	Very good	Medium green	Yes	Yes
Faverolles	Very good	Medium tinted	Yes	Yes
Fayoumi	Fair	Small white	No	No
Hamburg	Very good	Small white	Yes	No

Lay rate: Excellent (5 or 6 per week), very good (4 or 5 per week), good (3 or 4 per week), fair (2 or 3 per week), poor (1 or 2 per week)
Egg size: Extra small (1.25 oz or less), small (1.5 oz), medium (1.75 oz), large (2 oz), extra large (2.25 oz), jumbo (2.5 oz or bigger)
Cold hardy: Tolerates cold winter temperatures
Winter layer: Continues laying in winter

HEAT HARDY	FORAGER	BROODY	DOCILITY	CAUTIONS
Yes	Average	Frequent	3	
No	Better	Excessive	4	
No	Average	Frequent	3	Noisy
No	Average	Seldom	3	Not suited for wet weather
Yes	Better	Frequent	2	Both roosters and hens can be aggressive
Yes	Better	Occasional	4	
No	Better	Frequent	4	
No	Better	Frequent	4	
Yes	Average	Frequent	3	Roosters can be aggressive
Yes	Better	Infrequent	3	Roosters can be aggressive; will not breed true
Yes	Best	Seldom	1	
Yes	Better	Occasional	4	
Yes	Better	Occasional	4	Often get picked on in mixed flocks
Yes	Best	Seldom	1	
Yes	Best	Seldom	2	Flyers

(continued)

Heat hardy: Tolerates summer heat especially well
Forager: Able to hunt for food
Broody: Wants to hatch eggs
Docility: 5 (potential lap bird), 4 (friendly), 3 (average), 2 (flighty), 1 (avoids human contact)

BREED SELECTOR GUIDE *(CONT.)*

BREED	LAY RATE	EGG SIZE AND COLOR	COLD HARDY	WINTER LAYER
Holland	Good	Large white	Yes	No
Houdan	Fair	Small white	No	No
Jaerhon	Very good	Large white	No	Yes
Japanese Bantam	Poor	Extra small white	No	No
Java	Good	Large brown	Yes	No
Jersey Giant	Good	Large brown	Yes	Yes
Kraienkoppe	Fair	Medium tinted	No	Yes
La Fleche	Good	Large white	Yes	No
Lakenvelder	Good	Medium white	Yes	No
Langshan	Good	Medium brown	Yes	No
Legbar	Very good	Medium blue	Yes	No
Leghorn (Non-white)	Good	Large white	No	No
Leghorn (White)	Very good	Extra large or jumbo white	No	No
Malay	Poor	Medium brown	No	No
Marans	Good	Large chocolate	Yes	Yes

Lay rate: Excellent (5 or 6 per week), very good (4 or 5 per week), good (3 or 4 per week), fair (2 or 3 per week), poor (1 or 2 per week)
Egg size: Extra small (1.25 oz or less), small (1.5 oz), medium (1.75 oz), large (2 oz), extra large (2.25 oz), jumbo (2.5 oz or bigger)
Cold hardy: Tolerates cold winter temperatures
Winter layer: Continues laying in winter

HEAT HARDY	FORAGER	BROODY	DOCILITY	CAUTIONS
No	Better	Frequent	4	
No	Best	Frequent	3	
No	Better	Seldom	2	
No	Average	Frequent	3	Roosters can be aggressive; flyers
No	Better	Frequent	4	
No	Average	Occasional	3	Slow to mature
No	Best	Frequent	2	
No	Best	Seldom	1	
No	Better	Seldom	2	
No	Better	Occasional	3	
Yes	Best	Occasional	3	Sex-linked breed
Yes	Better	Seldom	2	Flighty; large combs especially subject to frostbite
Yes	Better	Seldom	2	Flighty; large combs especially subject to frostbite
Yes	Better	Occasional	1	Both roosters and hens can be aggressive
No	Better	Frequent	3	

(continued)

Heat hardy: Tolerates summer heat especially well
Forager: Able to hunt for food
Broody: Wants to hatch eggs
Docility: 5 (potential lap bird), 4 (friendly), 3 (average), 2 (flighty), 1 (avoids human contact)

BREED SELECTOR GUIDE *(CONT.)*

BREED	LAY RATE	EGG SIZE AND COLOR	COLD HARDY	WINTER LAYER
Minorca	Very good	Extra large white	No	No
Modern Game	Poor	Medium white	No	No
Naked Neck (Turken)	Fair	Medium tinted	Yes	No
New Hampshire Red	Good	Large brown	Yes	Yes
Old English Game	Fair	Medium white	No	Yes
Orpington	Very good	Large brown	Yes	Yes
Penedesenca	Good	Medium chocolate	No	No
Phoenix	Poor	Medium tinted	No	No
Plymouth Rock	Very good	Large brown	Yes	Yes
Polish	Fair	Small white	No	No
Redcap	Good	Medium white	No	Yes
Rhode Island	Excellent	Extra large brown	Yes	Yes
Rosecomb Bantam	Poor	Extra small tinted	Yes	No
Russian Orloff	Fair	Medium tinted	Yes	No

Lay rate: Excellent (5 or 6 per week), very good (4 or 5 per week), good (3 or 4 per week), fair (2 or 3 per week), poor (1 or 2 per week)

Egg size: Extra small (1.25 oz or less), small (1.5 oz), medium (1.75 oz), large (2 oz), extra large (2.25 oz), jumbo (2.5 oz or bigger)

Cold hardy: Tolerates cold winter temperatures

Winter layer: Continues laying in winter

HEAT HARDY	FORAGER	BROODY	DOCILITY	CAUTIONS
Yes	Best	Seldom	2	
Yes	Better	Occasional	3	Can be aggressive with other birds; intolerant of confinement
Yes	Better	Frequent	4	
Yes	Better	Seldom	4	Hens can be aggressive with other birds
Yes	Best	Excessive	3	Flyers; roosters can be aggressive
Yes	Better	Frequent	5	
Yes	Best	Seldom	1	
No	Average	Occasional	2	Require special housing to maintain long rooster tails
Yes	Better	Infrequent	4	
No	Average	Seldom	3	Often get picked on in mixed flocks
No	Best	Seldom	3	
Yes	Better	Seldom	4	Hens can be aggressive to other breeds; roosters can be aggressive to humans
Yes	Average	Seldom	3	Flyers; roosters can be aggressive
Yes	Better	Seldom	2	

(continued)

Heat hardy: Tolerates summer heat especially well
Forager: Able to hunt for food
Broody: Wants to hatch eggs
Docility: 5 (potential lap bird), 4 (friendly), 3 (average), 2 (flighty), 1 (avoids human contact)

BREED SELECTOR GUIDE *(CONT.)*

BREED	LAY RATE	EGG SIZE AND COLOR	COLD HARDY	WINTER LAYER
Sebright Bantam	Poor	Extra small white	No	No
Serama	Poor	Extra small tinted	No	No
Sex Link	Excellent	Extra large or jumbo Brown, mostly	Yes	Yes
Sicilian Buttercup	Fair	Small white	No	No
Silkie Bantam	Fair	Small tinted	Yes	Yes
Sultan	Poor	Small white	No	No
Sumatra	Poor	Medium white	No	No
Sussex	Very good	Large light brown	Yes	Yes
Welsummer	Very good	Large chocolate	Yes	Yes
White-Faced Black Spanish	Good	Large white	No	No
Wyandotte	Very good	Large brown	Yes	Yes
Yokohama	Poor	Small tinted	No	No

Lay rate: Excellent (5 or 6 per week), very good (4 or 5 per week), good (3 or 4 per week), fair (2 or 3 per week), poor (1 or 2 per week)
Egg size: Extra small (1.25 oz or less), small (1.5 oz), medium (1.75 oz), large (2 oz), extra large (2.25 oz), jumbo (2.5 oz or bigger)
Cold hardy: Tolerates cold winter temperatures
Winter layer: Continues laying in winter

HEAT HARDY	FORAGER	BROODY	DOCILITY	CAUTIONS
Yes	Average	Seldom	3	Chicks notoriously difficult to rear
No	Average	Occasional	3	Smallest chickens in the world (12 oz on average), but size can be variable based on the breeding line; they are usually too small for regular pelletized food
No	Average	Seldom	3	Will not breed true
Yes	Average	Seldom	1	Wild
Yes	Average	Excessive	5	Stubborn broodies; sometimes prefer to sleep on the floor rather than on roosts
No	Average	Seldom	4	
Yes	Better	Frequent	2	Males have multiple spurs
Yes	Better	Occasional	5	
Yes	Best	Occasional	4	
No	Better	Seldom	2	Face color takes time to develop
Yes	Better	Infrequent	3	Can be aggressive to other birds
No	No	Occasional	1	Require special housing for long tails; can be aggressive to other birds and to humans

Heat hardy: Tolerates summer heat especially well
Forager: Able to hunt for food
Broody: Wants to hatch eggs
Docility: 5 (potential lap bird), 4 (friendly), 3 (average), 2 (flighty), 1 (avoids human contact)

CHAPTER 4

TRUTH in (CHICKEN) ADVERTISING

Let's take a look at how chicken breeds are described in catalogs—
and how to decode what those descriptors mean. Here's where we'll
explain the terms used in the Breed Selector Guide that starts on page 18
as well as go over other considerations when "casting" your flock.

If you feel like you need some guidance on becoming a chicken parent and whether you'd do best to begin with eggs or chicks, the following descriptions of brood characteristics will give you an in-depth look at what to expect.

BROODINESS

Broody can be used to describe the behavior of a certain hen, but it can also describe the tendency of a whole breed. Broodiness is a hormonal condition. When *a particular hen* is broody, this means she wants to hatch her eggs. A *breed* that's described as broody has hens that often, individually, go broody. Some chicken breeds will go broody often and some are less

likely to ever go broody. Some will stay broody a long time even with nothing in the nest, while others will snap out of it quickly if they don't have eggs to set on. This can be a good thing or a bad thing, depending on what you're looking for.

If you plan to hatch fertile eggs at home, a good broody is the best incubator you can have. The power never fails, and the humidity is always right. It's the most natural way to raise chicks. See page 78 for a more detailed discussion of broody hens.

It can be charming when your hens are broody. Once a day or so they'll emerge from the nest like whirling dervishes: All their feathers will be ruffled, and they'll hold their wings away from their bodies to make themselves look even bigger. Sometimes they'll just sneak away from the nest for a bit with some persistent clucking, but more often they'll rise with a terrible screech and charge any hen that gets in the way. They are adorable.

But broodiness can also be a pain in the neck.

One issue is that hens stop laying eggs while they're broody. If you have several hens go broody at once, this can mean a big drop in the number of eggs you get—and a drop in the number of nests available to the rest of your flock.

If a broody hen is not hatching, it can still be problematic. She may go from nest to nest adopting the eggs other hens leave behind. She may actually steal them one at a time, rolling them along with her beak and gathering them together to set on. Again, this doesn't sound bad—except that some broodies will peck at you to keep you from checking for eggs in their nests.

Broodiness is more annoying to human caretakers than to the rest of the flock. There's the pecking. There's the fussing. Sometimes they won't go in the coop at night if they've gone broody outside. Broody breeds don't even need eggs to set on to be broody, and they won't know the difference between fertile and infertile eggs. They may go broody on golf balls, or old doorknobs, or even air. Because they eat little during brooding, they may lose weight and condition in addition to the cessation of laying. They may also pluck out their breast feathers so they can

The special way a broody hen clucks can encourage other hens to become broody, too. Here we see a quartet of broodies, happily sitting away.

be closer to the eggs—if there are any—and they may be ill-tempered and territorial.

If you have a broody hen and she is not hatching fertile eggs, try to lift her off the nest at least once (or even several times) during the day so she will be reminded to get enough to eat and drink. In addition, since broodies don't get off the eggs to dust bathe, for example, they are vulnerable to infestations of mites and lice.

Eventually, your broody will get over her broodiness and begin ranging with the rest of the flock again. After a while, she'll begin laying eggs again, too! But is this something you want to deal with when it comes to your flock?

Egg-cetera!

fresh eggs

If you are storing eggs for incubation, studies show they hatch best when they've been stored at about 60°F.

The bottom line: Having a flock full of frequently or excessively broody birds can often be more of a headache than not. If you will not be hatching at home or you don't want to worry about broodiness very often, consider choosing breeds listed as infrequently or seldom broody. If you're looking for top producers, you probably want to avoid excessively broody birds.

EGG-LAYING RATE

If you read breed descriptions in chick catalogs or online, you'll see some breeds listed as "excellent layers" or "productive layers." Others may be described as "fair layers" or "good layers." But what does this mean exactly? There are no standard terms used by hatcheries and breeders for egg-laying rates. For instance, from one hatchery, "good" might mean you can expect an average of three or four eggs a week per hen; from another, it might mean the average is six or seven. Sometimes lay-rate descriptors may not even be consistent within the same hatchery. If lay rate is important to you, the chart in Chapter 3 is standardized so the descriptors refer to specific rates; it's important to note that our rating system is not standard to hatcheries and breeders. In our chart, "excellent" means five or six eggs per week, "very good" means four or five, "good" means three or four, "fair" means two or three, and "poor" means one or two.

Remember, laying rate refers to an *average rate for mature birds*. Young birds may lay more, initially. Older birds will probably lay fewer. For instance, young Rhode Island Red hens that are excellent layers may lay an egg a day every day

for many months even though the average lay rate listed is only five or six a week. But a 6-year-old Rhode Island Red may lay only two or three eggs per week, or even none at all.

That's not all there is to it, though. Just because you choose a breed with an "excellent" laying rate doesn't necessarily mean that your hens will be laying lots of eggs. For instance, Silkies are excellent layers, but they go broody so often it's likely that they won't be laying for much of the year, since birds don't lay while they are brooding. Anconas are excellent layers, too, and they seldom go broody. However, they're poor winter layers, so if you live in an area that gets cold and dark during the winter, then for at least a quarter of the year (maybe more) you shouldn't expect much from them.

Consider this example: If you want two dozen eggs a week or so, you'd only have to keep about four Rhode Island Reds to hit that target number. But if you want to keep funky, crested Polish hens, you might need 12 of those—or more—to get the same number of eggs. And even then, the Polish eggs would be smaller than the Rhode Island Red eggs, so you'd have to use more of them in your recipes.

Egg-laying rate is often an important consideration if you live in a place where the number of birds you can legally keep is limited to a very small number, such as three or four. Clearly, an unfortunate side effect of such a restriction is that it also limits which breeds you can keep if you hope to gather enough eggs to meet your family's needs. And, honestly, it's easy to get caught up in choosing the most beautiful and unusual breeds, only to realize some months

If your family is big on eggs, consider that four or five young hens will typically provide you with only three eggs a day!

later that you overlooked how few eggs your favorite breeds would be giving you.

In how-to books and magazine articles about chicken keeping, it's common to read that a small family needs four or five hens to meet its egg requirements. And that *can* be true, sure. And it's probably mostly true for average-size families (three to five members) and excellent layers that are in their prime. However, we're not always talking about average-size families or excellent layers or hens in their prime. Most families that keep a small flock of backyard birds don't kill those pets as they get older and lay fewer eggs. And many prefer to get unusual breeds: breeds that lay blue or green eggs, breeds with personalities more like kittens or Tribbles than chickens, breeds that are only as large as pigeons, breeds that have huge crests of feathers or beards or feathered legs or extra toes.

Finally, it is important to note: Whether a hen lays up to the potential of her breed also depends on the care you provide. Stress, improper feed, and illness can all reduce laying. Laying will also slow down—and may temporarily cease—during the annual molt. If you live in an area with excessive heat, your birds may see reduced laying in the worst of summer, too. And if your birds need lots of space and they are kept in a confined run, they will be stressed and not lay well. So, remember, when casting your flock, you'll want to choose hens that do well in the conditions they'll experience in your care.

EGG SIZE

Size is also an important consideration if you love eating eggs. When casting your flock, keep in mind that if you get six eggs a week from a bird that lays tiny eggs, obviously it's not going to be the same as getting six eggs a week from a bird that lays extra large or jumbo eggs. So the number of eggs per week is not the be-all and end-all of determining if you will be getting as many eggs from your hens as you'd like to have. Read more about cooking with different sizes of eggs (and see egg-size classifications) in Chapter 13.

EGG COLOR

For many of us, half the fun of keeping chickens is marveling at the beautiful and unusual eggshell colors they produce.

You may have heard that certain colored eggs are healthier to eat than others. That's not true, although it's a common myth. When it comes to eggs of different colors, there is no difference in terms of edibility, healthfulness, or nutrition. The only thing that affects the nutritional value of your hen's eggs is what she eats. This means that if your hens have access to pasture and the opportunity to forage, their eggs will be more nutritious; studies have shown that hens with pasture access produce eggs that are lower in cholesterol and saturated fat, and higher in vitamins E, D, and A; beta-carotene; and omega-3s. As to why hens lay eggs with different colored shells, no one really knows for sure, although there are a few guesses. It's intuitive to think that in the wild the colors and patterns of bird eggs might help provide camouflage for them while they're in the nest, and there does seem to be a relationship between egg color and nest type in wild birds that suggests the difference has to do with

Many chicken keepers delight in getting eggs of every color and size.

camouflage—or perhaps it once had to do with camouflage. White eggs are chiefly laid by cavity nesters, where the eggs will be more concealed. Gray and brown eggs are more often laid by ground-nesting birds. Speckled eggs or blue eggs are often laid in open nests. There are a few exceptions, though. For instance, hummingbirds lay white eggs in an open nest.

However, definitively demonstrating that relationship has proved problematic. For instance, in one experiment, a scientist painted eggs different colors (brown, white, blue, spotted) and placed them in nests with varying degrees of cover, then kept track of which eggs suffered predation. The color didn't seem to have any relationship to the predation suffered, and additional experiments testing the effectiveness of egg camouflage have been repeated a few times with similar results.

So why do chickens lay different colored eggs? The best we can tell you for sure is, they just do.

When it comes to the mechanism that creates the color, that's a little easier to explain. Eggshells start out white; that's the natural color of the substance that makes up eggshells. Originally it was thought that the color pigments were synthesized in the hen's blood by the breakdown of hemoglobin, but research shows it is more likely that they are actually synthesized within the shell gland pouch. The blue color is produced by biliverdin, and the brown color is produced by protoporphyrin. These "dyes" are incorporated into the shell in varying ways based on their different compositions.

When eggs are brown, the color is sort of "painted" onto the white egg inside the chicken's reproductive tract by the shell gland pouch. (When you crack brown eggs, they are brown on the outside and white beneath.) As you know, some brown eggs are very light in color, while other breeds lay extremely dark chocolate-brown eggs. To simplify this a little, the darker layers just produce more of the brown color to go on the shell. There are variations, of course. Faverolles, for example, tend to lay eggs that are pinkish brown; other breeds lay eggs that are more of a golden brown. However, the brown "paint" all goes on the same way.

The mechanism for creating blue and green shells is a little different. For blue eggs, the blue color actually goes all the way through the shell, even to the inside of it. (In other words, when you crack blue eggs, they'll be blue on the inside, too.) Again, the brown color can be thought of as being painted on the outside, while the blue color is distributed throughout the shell.

Green eggs are laid by chickens that have both blue and brown egg-laying genes. The blue ends up throughout the shell as described above, and the brown is painted on top, creating a green appearance. (When you crack these shells, they'll be green on the outside and blue on the inside.) Because green eggs have both pigment types, they can be exceptionally hard to candle; *candling* refers to shining a light through the eggshell to see what's inside. Candling is harder sometimes with green eggs— even light green eggs—than it is with deep chocolate-brown eggs.

Interestingly, while individual hens lay basically the same color throughout their lives, they lay their darkest eggs at the beginning of the season in the spring, after their bodies have had a winter break. When they hit their egg-laying stride in the summer and they're laying as quickly as they can for their breed, often their

Egg-cetera!

fresh eggs

Brown eggshell colors are deposited on the outside of the shell, while blue goes throughout the shell, even on the inside.

eggs lighten up in color. Heat stress can also cause eggs to lighten. (Some illnesses may cause this, too.) As chickens lay at a faster rate and their eggs get larger, the eggs will lighten because there is only a certain amount of color each hen produces. If color is laid over more eggs and/or over a larger surface area, the "paint" will be thinner, obviously.

This is important to know if you're choosing breeds especially for egg color. If you want only chocolate layers, for instance, remember that their egg color will naturally vary in intensity over the course of the season. The eggs of Marans can go from chocolate to a color that's not much darker than the egg of a regular brown layer. After a break in laying, such as a molt or a period of broodiness, their eggs will become darker again.

COLD HARDINESS

This seems pretty straightforward, doesn't it? But if you see a breed described as "cold hardy," there are a few things that might not be obvious from the get-go. First of all, be aware that cold hardiness and heat hardiness are not mutually exclusive. Some breeds tolerate both extremes, while others don't do especially well in either. Next, know that just because a chicken is cold hardy doesn't necessarily mean she is a good winter layer. For instance, Easter Eggers are cold hardy birds, but after their first winter, they are notoriously terrible winter layers. In fact, most breeds don't lay particularly well in the short days and cooler temperatures of winter.

Another thing to consider is that, given

This White Plymoth Rock is a good cold hardy choice.

proper care, nearly all breeds will do well in most areas of the United States. However, in some parts of Alaska, Minnesota, North Dakota, and northern New England (and in high elevations), you'll want to be sure to get cold hardy breeds only.

A lot of times, people will write us at My Pet Chicken to ask, "How cold can they get?" We understand and appreciate the concern, but this is not necessarily a question that can be answered. Think about it: Is there a minimum temperature for your dog or cat? Or for you? No. And, frankly, we're glad that we don't know of a study where scientists are testing to see how cold a chicken (or a dog or a cat or a human) can get before it dies. Animals are not like peach trees, which have specific USDA-hardiness zones. Plus, how much the cold affects chickens will depend on how long your chickens are exposed, what kind of shelter they have, whether they're out of winter winds, how well fed they are, whether it's wet or dry, and other factors. Please turn to page 66 for helpful information about chicken coop options.

WINTER LAYER

As mentioned earlier, cold hardiness and winter laying capability do not equate. Also, just because a breed is a winter layer doesn't mean the hen will continue laying at top capacity all season. She is still likely to take a break during the molt, and when she starts again, she won't be laying as many eggs until the days get longer. That said, winter laying is one of those qualities for which there are no real downsides, as there are for broodiness, for example. And it's probably going to be especially important for very small flocks. Casting good winter layers in your flock is helpful.

HEAT HARDINESS

The same considerations listed under "Cold Hardiness" apply to heat hardiness as well. Heat hardiness and cold hardiness are not mutually exclusive, and even birds that are heat hardy may not lay well in periods of sweltering heat. Extreme heat is actually more difficult for most breeds to bear than extreme cold, but, even so, remember that you'll only need to seek out heat-hardy breeds if you live in areas where there are extended periods of extremely hot temperatures, like the desert areas of the Southwest or the Deep South. And, finally, there is no precise "how hot can they get" temperature, just as there

is no "how cold can they get" temperature. How much the heat affects your chickens will depend on such variables as how long your chickens are exposed, what kind of shelter they have, whether they have shade and access to cool water, and whether it's moist heat or dry heat.

FORAGER

All chickens forage. It's what they do. However, some breeds are better at it than others and instinctually range farther, have a quicker eye, and are better at catching bugs. Others, bless their feathery little hearts, tend to stay closer to home and just aren't as good at finding wild good-to-eats. Some breeds, like the Buckeye, are known as particularly avid hunters and good mousers.

You read that right.

Yes. Chickens will eat mice, moles, voles, shrews, small snakes, and other creatures. They are voracious omnivores. Think of them as tiny dinosaurs, smaller than velociraptors. The bottom line is that they're hunters, though, even if what they hunt is on the small side.

Good foragers will be a special help in reducing the bug population around the coop. For example, there are a lot of ticks in my area, but I never see any around the house because the chickens more or less clean them out. If you live in an urban or suburban area, though, and your chickens will be confined to a small run or yard, don't focus on choosing breeds with especially good foraging abilities. It may be a waste of time, at best, and, at worst, you'll have trouble keeping the flock out of your landscaping.

DOCILITY

When looking through breed descriptions and hatchery catalogs, you'll see lots of different descriptors. Some common ones include *friendly, flighty, calm, wild, active, alert,* and *tolerant/intolerant of confinement*. For the most part, these adjectives are self-explanatory. They're also quite subjective.

If a breeder was lucky enough to have a tame line of Penedesencas (this breed is usually on the flighty side, avoiding human contact) and concluded that they are friendly, it's not that the breeder is wrong, it's just that she's had a different experience than most. In the chart in Chapter 3, we relied on a number of sources, including the experiences of various breeders and hatcheries as well as our own.

People with little experience with chickens get confused when they see conflicting reports, or when they get a chicken that was supposed to be friendly, and isn't. They seem to regard these adjectives more as Rules Chickens Must Follow than as general observations about the personalities of different breeds. Just because a certain breed of chicken has a reputation for friendliness doesn't mean that *every* bird of that breed is guaranteed to be friendly, the same way *every* Golden Retriever is not going to be friendly.

The word *flighty* can cause particular confusion, too. *Flighty* generally means nervous—and a nervous chicken can be easily startled into taking flight. Of course, all chickens can fly. It's just that some breeds are more prone to flying, and some can fly higher than others. Birds described as *flighty* or *flyers* will generally require higher fences to keep them confined, and they may be on the nervous side—or not. Ameraucanas are flyers, but they're so friendly and forward that they're more prone to flying toward you than away from you. Still, if you're not keen on having a chicken perching, and

Lissa's Scratchings

The Orpington has the reputation of being a "lap chicken." My Orpington has been docile and calm, but not particularly friendly. Most of my Easter Eggers beg to be petted, while two are rather wild and wary of me. My Silkies have all been like kittens—except for one who seemed to have a perpetual anger-management problem. One of my Cuckoo Marans was a calm love-bunny, while another was standoffish. Sussex chickens in general don't have a reputation for above-average friendliness, yet almost all the My Pet Chicken employees have had Sussex hens that were especially loving, almost like puppies. For this reason, they were given the highest rating for docility in the Breed Selector Guide on page 18.

possibly pooping, on your shoulder or on top of your head, you may want to avoid flyers. Another consideration is that small flyers, especially Bantams, may try to roost outdoors in trees, but it may not always be safe for them to do so. And trying to herd tree-roosters back into the safety of the coop at night may be one of the goofiest things you'll ever have to do.

You will not want to consider any single catalog, book, or web site as definitive in the matter of breed personalities. Start with this book but explore several sources, when possible. If you see a source that disagrees with your preconceived notions or with other opinions you've read before, don't immediately consider it wrong, just keep it in mind and let it help inform your opinions as you continue to research which breeds might work for you.

Remember that the "specs" for chicken breeds are not objective specifications like you would find for tech gear (one computer may have exactly 300 GB of memory, 4 GB of RAM, and a monitor with a specific quantifiable resolution). Instead, chicken specs are subjective descriptions of living creatures. If you do extensive research and find that 95 percent of your sources are at a consensus, that will tell you one thing; if you find that there is substantial diversity of opinion, that will tell you something else.

FANCY FEATHERING

There are some other factors to be mindful of when it comes to casting your flock. For instance, you may be drawn to unusual-looking birds with fancy feathering. Some chickens have feathered legs, large crests, ear tufts, muffs, and/or beards of feathers. There are some drawbacks to fancy feathering, though, so consider carefully before you make any decisions.

For example, birds with fancy feathers can be vulnerable to getting picked on in a mixed flock. This isn't always the case—beards don't seem to draw as much attention as ear tufts, crests, or feathered legs—but it happens frequently enough. Lissa had to give away a Rhode Island Red—a great layer in the prime of her life—because she just couldn't stop picking at the leg feathers of other birds. She would pick them bloody. All her other Rhode Island Reds got along fine with her feather-legged birds. There were better things to forage, as far as they were concerned. But this one particular bird was just driven to pluck leg feathers. Rather than invest in weekly treatments of an antipicking lotion to spread on her feather-legged birds, Lissa found a new home for the hen in a flock that didn't include any feather-legged birds.

Crested birds face a particular danger: Their

Egg-cetera!

Fertilized eggs are perfectly fine to eat. It is impossible to taste the difference, and—with the naked eye—it is barely possible to see the difference between fresh fertilized and unfertilized chicken eggs.

large head crests can obscure their vision, and, as a result, they may not see an attack coming, whether it's from another chicken wanting to pluck at crest feathers, or a predator like a dog, raccoon, or hawk. Because they are more vulnerable to predators, they tend to fall lower in the pecking order than other breeds. In addition, crested birds sometimes have a hard time seeing how to get back into the coop at night. Granted, it's adorable when a little puff of a Silkie needs to be carried into the coop at night because she's so covered in fluff that she can't see, but it can get tedious. That said, some birds with especially large crests may need regular trimming to improve their vision. (If you're planning to exhibit your birds at shows, of course, you will not want to trim feathers.) Also, in the wintertime, if a crest gets wet while the bird is drinking, it may freeze, creating the risk of frostbite.

When it comes to rough weather, birds with feathered legs face potential problems as well. In slushy or muddy cold, the leg feathers of birds with profusely feathered legs can get caked, then freeze hard. Even in warm weather, leg feathering can create difficulties in wet conditions. When leg feathers get covered in mud, all your eggs may get dirty as mud is carried into the nests.

MIXING BREEDS

So, let's get back to being a casting director for your flock: Don't approach this like you're casting for reality TV. You don't want drama. You're not going for sensationalism that will draw big Nielsen ratings, right?

While most hens will get along well in a mixed flock, there are circumstances where you don't want to keep mixed breeds. The truth is, there are some breeds that may not do as well as

There's nothing quite like a flock flaunting a variety of colors and feather patterns, but sometimes differences can be dangerous. When mixing breeds, provide plenty of room to roam and range.

others in a mixed flock. You usually don't want to pair breeds with an aggressive reputation and breeds that tend to be very submissive, since the submissive birds can get picked on. You don't want that sort of reality-show strife in your flock every day!

Likewise, if you are mixing breeds in your flock—and most backyard chicken keepers like to do this—make sure you are mixing birds somewhat equally. This doesn't mean you have to have the exact same numbers of each breed, but it does mean that you don't want to get five Wyandottes and pair them with one solitary Polish with her huge and goofy crest. They say that "birds of a feather flock together," and it's true! What this means in terms of fashioning your own flock is that you don't want to create a situation in which there is no one for some of your birds to flock with. Everyone needs a like-minded friend. Chickens who look very different from the rest of your flock can get picked on, and, as mentioned before, birds with fancy feathering sometimes suffer for it more than others because they may not appear to be chickens to the other flock members.

In fact, the most problematic birds to include in a mixed flock are Polish or other crested birds, because large crests can prevent them from seeing an attack coming, and Faverolles, because they can be so submissive that they usually end up at the bottom of the pecking order. However, if you are raising a flock of many chickens that all have different looks, your birds will have a much broader idea of what a chicken should look like and won't usually pick on one another. Of course, the more room they have, too, the better they will be. If your flock can range on pasture every day, they have things to occupy their thoughts other than why Bessie is wearing that silly and unusual plumage!

Even though there are some important considerations when creating a mixed flock, in most cases you can keep a variety of breeds together without trouble. You can even include both bantam and large fowl breeds in your flock, if that is what you prefer. In fact, contrary to what you might think, bantams don't always—or even usually—end up on the lower end of the pecking order just because they're smaller. When mixing birds of different sizes, you will simply need to make certain that feeders and waterers are set at a height that all your chickens can easily reach.

So, remember, when you're making casting decisions for your mixed flock, be thoughtful and deliberate, and understand that while you have some control, you can't control everything. No one is going to be able to look at your prospective list of birds and give you absolute assurance that they'll all get along. This applies even if you are interested in keeping only one breed. Some chickens will be insufferable jerks while others will be pitiable doormats—just like humans.

BECOMING A CHICKEN PARENT

Before you embark on your chicken-keeping adventure, you'll need to decide whether to start by incubating fertile hatching eggs, purchasing baby chicks from a mail-order hatchery, buying juveniles, or adopting battery hens. Each option has its pros and cons, and your choice will be based on how much time (and risk tolerance) you have.

Starting with Eggs, Chicks, Juvenile Chickens, or Adult Birds

This is really important, so please pay attention: *In almost every case, starting a laying flock with baby chicks will be easier and less expensive than starting one by hatching eggs at home.*

Got it?

Let us explain why. Don't get us wrong, home hatching is great! Personally, we love hatching at home, and there's nothing like seeing baby chicks pop out of their eggs and take their first wobbly little steps. We love that they don't have to ship anywhere; they can go straight to their brooder. When we hatch with broody hens—which is as often as possible—we love seeing the mother hen with her chicks. There's nothing better. But for people new to the hobby, hatching eggs can be way more frustrating and expensive than it's worth. Incubation is not an easy thing to do.

You might expect that if you want eight hens, you need eight eggs. Nope. If you want eight hens, at minimum, you'll probably want to start with 32 eggs—or more. And even then, results are not guaranteed. The problem is that fertile eggs (supplied from any source) are not guaranteed to hatch. If you're starting from scratch, you'll probably have to get fertile eggs shipped to you, and the truth is that shipping is just very rough on fertile eggs. With shipped eggs, the average hatch rate is about 50 percent. That's just the *average*;

it's not a guaranteed rate. You can have higher or lower hatch rates. Sometimes, much higher or much lower. In fact, with fertile eggs, you may do everything right—and the egg provider may do everything right—and you still might not have any eggs hatch. Traveling by mail, your eggs may see pressure changes in flight, a sweltering truck or an icy warehouse, bumpy roads, or thoughtless handling. Even if you're lucky enough to get fertile eggs right next door—eggs that have never shipped—not all of them will hatch.

Plus, about 50 percent of what hatches will be male, on average. And, again, that's just the average!

Getting low hatch rates every so often is just the nature of the hatching beast. *If you are a beginner and want a laying flock, almost every time, it's better to start with baby chicks rather than fertile eggs.*

Buying fertile eggs is appropriate when you are looking for a rare breed or color that you just can't find as a baby chick, when you are hoping to have several good roosters to choose from for your home flock, or when you want the experience of hatching eggs—in your home incubator or under a favorite broody. Even then, though, you'll have to be mentally prepared for a terrible result: one hen—or possibly none—for all the effort and expense.

If you want to start with hatching eggs, you need to be okay with throwing that money away if you get less than desirable results. So if you have to save for months to buy those dozen fertile eggs, don't do it! Starting with hatching eggs is not the way to go for you. This is because— unlike starting with baby chicks, where the

These irresistible Blue Ameraucanas (left and center) and Splash Ameraucana (right) chicks are just hours old.

hatchery usually makes some guarantees about what you'll end up with—*you're* taking all the risks when you start with fertile eggs. If you are set on starting with fertile eggs, then your budget should be large enough so that you can simply shrug your shoulders and move on if your hatch doesn't turn out as expected.

Order hatching eggs only if you are prepared to be very flexible with whatever your results may be.

WHAT ABOUT JUVENILE CHICKENS?

Some people want to get started with juvenile "point-of-lay" chickens, often called started pullets. Pullets are young female chickens, not yet of laying age. These are birds that have been raised to 12 to 16 weeks old. Juvenile birds have their advantages, the main one being that you don't have to wait as long for your flock to begin producing eggs. In addition, presuming you're buying from a reputable source, there won't be any sexing errors to deal with. With very few exceptions, juvenile birds can be reliably sexed. With baby chicks, there is always the possibility of a vent-sexing error; vent sexing is a method of determining sex that involves looking at a day-old chick's vent (the opening beneath its tail,

Egg-cetera!

fresh eggs

A hen who is laying well will have a moist, pink vent and a wide pelvis. Her legs may also be paler or bleached (this occurs in brown egg–laying breeds with yellow legs). Good layers may also have broken feathers, because their bodies put fewer resources into plumage.

Lissa's Scratchings

Hatching eggs is a gamble. Believe me, I know from personal experience that hatching eggs is chancy. When I was getting started, I decided to use fertile eggs and one of the cheap Styrofoam incubators. And I was smart about it—my expectations were reasonable. I incubated 18 eggs with the hope of four females or so at the end. My math was this: Nine or so chicks will hatch, and half or so of those will be female, so I should end up with four or five laying hens, right? Odds are I'll have at least three: That's basic math! I felt reasonably sure that everything would work out as I'd planned.

But here's what really happened: Eight out of the 18 hatched, which is not bad. Close to average, and certainly within any margin of error. On average, I should still expect four females, or so I thought—five or maybe even more if I was lucky! But as it turned out, seven out of the eight chicks ended up being males.

Out of 18 eggs, I hatched only one female for my laying flock—and one hen does not make a good laying flock!

I went in thinking that I would save money by starting with hatching eggs. After all, eggs are less expensive to ship than chicks, and the eggs I chose were also less expensive than chicks. (That's not always the case!) In my case, though, I paid about $65 for my eggs. If I had ordered female day-old chicks of the breed I wanted, it would have cost more than $100 because I would have had to order 25 chicks (25 was the minimum number of chicks most hatcheries would ship in the days before My Pet Chicken was around). Clearly I would save about $40 by starting with eggs.

But I was new to chickens, and I hadn't considered everything. I overlooked a few very simple things.

What I failed to consider was that if I started with chicks, I would be virtually guaranteed to have my laying flock at the end of the process. Even if by some crazy mischance there were sexing or packing errors and I ended up with only one hen (as I did with my eggs)—and I can't tell you how unlikely this is—I would at least get a refund for any errors. If there were numerous errors, I might qualify for a reshipment. If I got all 25 female chicks as ordered, I could even sell my extra baby hens, or give them away to the local 4-H club. I didn't think of that at the time.

Instead, with fertile eggs I wasn't guaranteed anything. I wasn't guaranteed they would hatch, much less how many females I would end up with. Plus, finding homes for any extra males I might hatch would be much more difficult than rehoming extra hens. And, in the end, I hatched just the one hen.

where both eggs and droppings exit). Another advantage: Juvenile birds don't need a special brooder like chicks do; they can be put in the "big girl" coop immediately. They're an attractive option for people who prefer not to care for baby chicks, or who just don't want to have to wait as long for eggs.

There are some drawbacks to beginning with started pullets, though. Aside from the whole you-don't-get-to-enjoy-them-as-chicks thing, started pullets are tremendously more expensive. Two birds may cost $100 or more, shipped. So unless you're looking for the smallest of flocks, this isn't an economical choice for most families.

In addition, with started birds, the breeds you have to choose from are usually severely limited. Most hatcheries offer only a few high-production breeds, such as Sex Links, Leghorns, and Rhode Island Reds, as started pullets. And those hatcheries will painfully "beak trim" their birds.

My Pet Chicken is the exception. We offer a good selection of heritage and unusual breeds—we've had Speckled Sussex, Blue Ameraucanas, Black Copper Marans, Barnevelders, and Javas, for example—and we do *not* beak trim. Even so, our juveniles are expensive when compared to the cost of baby chicks, and they are only available a few times a year.

You might be able to acquire juvenile chickens from a local farmer or breeder. Be sure you're getting chickens from a facility that follows the USDA's National Poultry Improvement Plan (NPIP). Buying poultry at auction is *not* recommended. Even if the birds you buy come to the event clean and healthy, they commonly pick up illnesses from other birds or animals while there.

THE RESCUE OF BATTERY HENS

Finally, you could consider rescuing some battery hens, meaning hens that have been confined to battery cages in factory farms for the duration of their lives. While you would be getting older hens, they generally still lay quite well, in backyard terms. They're also usually inexpensive, although you normally have to pick them up on-site. Plus, you know, there's the whole bonus of knowing you're doing a compassionate deed. But if you take this option, you may have illnesses and bad behaviors to deal with. The birds you get may have few feathers and require special attention. They may not know how to walk around, having been kept their entire lives in a place where they couldn't stretch their wings, much less walk. They may not be able to graze, because their beaks will have been trimmed. If you rescue battery hens, you'll need to be prepared to provide some extra TLC—and you'll definitely want to have access to an avian vet.

So, to repeat, starting with baby chicks is the easiest, most economical way for most people to start their home flock.

THE EASE OF BABY CHICKS

As we have said repeatedly throughout this chapter, starting with baby chicks is usually the best way to begin backyard chicken keeping. Chicks are relatively inexpensive to buy—and, if necessary, to ship. Starting with baby chicks means you can train them to be friendly from the get-go. You can choose whether you'd like to receive females or males, and sexed chicks will arrive with a good degree of accuracy.

Straight-run birds are an "as hatched" mix of males and females. Therefore, if you were to order eight straight-run birds, for example, you would get between zero and eight roosters and between zero and eight hens—there is no way to tell.

When you order straight-run chicks from a hatchery, they are not sexed first and then separated out into each order. The reason is that expert vent sexers of day-old chicks command top dollar (up to $2,000 per day!) since it is such a specialized skill. This is why ordering all straight-run birds is less expensive than ordering all females.

Generally, the hatching ratio of straight-run orders averages out to be a 50-50 mix, but you are more likely to hit the average if you have ordered large numbers. For instance, you would expect that if you flipped a coin 1,000 times, you would probably get heads around 500 times. Even if you had a few runs of flipping tails all in a row, things would equal out eventually by runs of heads.

If you flipped a coin only 10 times, though, there is no guarantee you would get exactly five heads and five tails. You might get four and six, seven and three, eight and two, or nine and one. It is even possible to get all heads or all tails, though it's not terribly likely.

You should only order straight run when you will be happy with any ratio, because if the chicks hatch with an out-of-whack ratio, you will receive an out-of-whack ratio. Unfortunately, it is thought that pullets are preferentially killed by variations in incubation temperature, while males are preferentially killed by cooler storage temperatures before incubation—so getting off-ratios does happen. Frequently. And, believe us, hatcheries wish they could control the number of males and females hatched out as much as you do!

With large fowl breeds, we think it pays to go ahead and just order the sexes you want. It isn't that much more expensive for a guarantee of pullets/cockerels. My Pet Chicken is currently the only hatchery in the country offering sexed bantams. Sexed bantams are more expensive, but if you are ordering just a few, having an off-ratio can seriously affect your flock. For instance, if you order five straight-run bantams, you would normally expect to get two or three cockerels, based on a 50-50 average. But if you got just *one* more cockerel than the average in

Egg-cetera!

fresh eggs

On average, pullets start laying eggs at about 6 months of age, depending on the breed. However, if your birds come into maturity at the coldest, darkest part of the year, they will sometimes not begin laying until spring—6 months is just the average.

These day-old chicks are still at the hatchery, waiting to be packed and shipped.

Where to Get Your Chicks

The two most common (and usually easiest) ways to acquire baby chicks, especially for those in urban or suburban areas, would be to purchase them from a mail-order hatchery or a farm/feed store. You might also choose to purchase some from a private breeder or farm.

HATCHERIES

If you want a flock that includes a few different rare or unusual breeds, a hatchery is usually your best choice. Every hatchery we know of has guarantees for live arrival and sexing accuracy. While problems can still occur, it will be your hatchery managing the losses, not you. Was there a bad hatch? Well, your order may get postponed or you may receive substitute breeds, depending on your hatchery and, when available, how you've asked them to deal with any issues with your order. But if you got a bad hatch from eggs you hatched on your own, it's just tough luck for you. You'd have to buy more eggs and try again—still with no guarantees.

Major hatcheries also go to great lengths to make sure the birds you receive will be healthy. They participate in NPIP and often have their flocks monitored for other illnesses as well. They have a wide variety of breeds, and they are good at estimating how many of each will be available each week. Unlike small breeders, they have the resources to pay an expert to sex the baby chicks.

They don't have magic wands, though. You can still receive sexing errors. Losses can happen. Hatch day issues and off-ratios can occur.

your straight-run order, that would put you at three or four cockerels out of five. Only one hen. And that would be—pardon our French—sucky. Very, very sucky indeed.

If you were to order, say, 50 straight-run birds, having a male-female ratio that is very askew is less likely to happen, so straight run might be something to consider with a larger order. Maybe. But it's not something we would recommend.

The bottom line is that if you want a small backyard flock, unless you're a big risk taker and willing to deal with the possible disappointment of getting fewer hens than you wanted, you'll probably be much happier receiving sexed birds.

But where to get them?

But getting your chicks from a hatchery is usually the best bang for your buck, and the least financially risky way to acquire chicks.

FEED STORES

Feed stores are another good way to get baby chicks. You will have fewer choices when you get chicks from a feed store, so there are a few limitations and cautions. But if this is an option in your area, it's something to consider.

Feed stores typically buy their chicks from hatcheries, so you'll have similar biosecurity protections since the chicks will most likely originate from NPIP flocks. Some feed stores order all vaccinated birds, while others don't order vaccinations at all. Is this something you want? Don't want? Be sure to ask. Another issue is that sometimes brooders are not kept clean in the store, whereas other times they are meticulous. A lot will depend on your local store.

Additionally, most feed stores offer only a limited selection of breeds. Often they order the same few breeds year after year, and they are generally high-production breeds like Leghorns or perennial favorites like Rhode Island Reds and Barred Rocks. Not that there's anything wrong with that—they're popular for a reason! Still, if you're looking for something different, it can be disappointing. You want hens that will lay chocolate-colored eggs? Or funny-looking Polish hens? Or fuzzy Silkies? You may have some difficulty finding them at your local feed store.

Another concern is that some feed stores order straight-run chicks rather than females. Sometimes feed store employees know a great deal about what they have ordered, and sometimes they don't. It's not always easy to tell who can be helpful, either.

Finally, keep in mind that even when you have knowledgeable, helpful feed store employees who provide the very best brooder conditions for their birds, which meet all the criteria you desire, the stores can still be thwarted by their own customers. Customers may pick up chicks from one brooder (for males) and inadvertently put them down in another (for females)—so this introduces another possibility for error.

Because the feed store will probably be placing your order among many other chick orders, you may end up saving money. It's less expensive *per chick* when more are shipped at a time. So while chicks at a feed store are usually a little more expensive if you're looking at just the cost per chick, that's because they usually already have shipping built into the price. It's not always the case that it will be less expensive, but it's certainly worth a little investigation.

Egg-cetera! *fresh eggs*

Even blind chickens are sensitive to day length. Chickens absorb light through their skulls, not just their eyes, and that triggers cycles of laying and not laying.

If you have children, consider breeds that tend to be docile, like the Easter Egger pictured here.

BREEDERS

When you're just getting started, local breeders are probably not going to be the best place to acquire birds. First of all, not all breeders are NPIP participants, which means they could be carriers for disease. Their birds may be more expensive, too, because it costs a small producer more to care for birds.

In addition, breeders usually cannot afford sexing services, so all orders may be straight run. They may not be able to provide vaccination services, either, if that's something you're interested in. Breeders usually keep only a few different varieties. Still, if those are the varieties you're looking for, who cares if they don't carry 60 different breeds! And it can be nice to support the local economy.

Just be careful to weigh the services your breeder can provide against those they can't provide to make an informed choice.

Stay Smart When Planning Your Flock

What it comes down to is *you*—what do you want in your flock? What's more important to you and what's less important? Choose wisely. But don't worry yourself to death about it, either. If you've genuinely tried to choose the best cast for your flock, it's pretty unlikely you've made any terribly dire mistakes that will destroy your happiness for the foreseeable future, right? Your flock is not going to explode in a fiery inferno at the slightest provocation, like a vehicle in a 1970s made-for-TV-movie car chase. Your flock will probably be less like a *Real Housewives* show and more like an episode of the *Golden Girls,* with each girl having her role to fill in the social structure and any argument softened by the joy of shared cheesecake. I mean, mealworms.

Flock Dynamic and Culture

Chickens are surprisingly bright. The chatter they make is not simply meaningless sounds; instead, it is a way of communicating with each other. Chickens are social creatures that naturally organize themselves into small communities called flocks. The flocks are organized by a social hierarchy called the pecking order.

Birds at the top of the pecking order have dominance over those lower in the order. It sounds cruel in a way; it's like a feudal social system, where everything belongs to the dominant birds by right, while lower-ranking, more submissive birds are only allowed what the highest-ranked flock members allow them to have. Even so, it's usually a peaceful arrangement. After all, chickens don't stockpile wealth like humans do. There won't be chickens sitting on piles of grain in your coop, some with too much to eat while others starve.

Instead, the highest-ranking chickens will eat and drink first. They'll get their choice of nests, nesting time, and roosting spaces. But *as long as there is enough space, food, and water for everyone*, even the lowest-ranking hens are going to be pretty contented. They'll just wait their turn. It doesn't matter that someone is at the top and someone is at the bottom; just make sure there is enough to go around.

According to Chris Evans, PhD, professor of psychology at Macquarie University in Australia, "Chickens exist in stable social groups. They can recognize each other by their facial features. They have 24 distinct cries that communicate a wealth of information to one another, including separate alarm calls depending on whether a predator is traveling by land or sea. They are good at solving problems." He notes that, as a trick at conferences, he sometimes lists these attributes, without mentioning chickens, and people think he's talking about monkeys.

As a keeper of backyard chickens, you'll find that one of the most interesting things about the hobby is the way the chickens interact with each other. You'll watch the way your flock socializes, and if you're observant enough, you'll start to understand "chickenese," and you'll even begin to recognize voices. Particularly if you keep more than one breed, you may notice that each breed even has its own typical voice.

Faverolles have a distinctive, pretty warble, and tend to be chatty with soft voices. Rhode Island Reds seem to say very little except when

What You May Not Know about Acquiring Your Chickens

Whatever method you choose, *be sure* the chicks come from an NPIP facility. NPIP stands for National Poultry Improvement Plan, and those that participate in this plan have certifications showing their flocks are monitored for some of the worst communicable illnesses chickens can get. My Pet Chicken and other major hatcheries are NPIP participants; most farm and feed stores purchase chicks from NPIP hatcheries. Some breeders, although not all, participate in NPIP—just check to be sure. You don't want to start out your hobby having to fight a contagious illness! And keep in mind that there are varying levels of NPIP certification; just because a breeder is certified by NPIP, it's not a promise that their birds aren't carriers for other types of illness. Some NPIP-certified breeders are only monitored for avian influenza and pullorum-typhoid; others have additional certifications for *Mycoplasma gallisepticum*, *Salmonella enteritidis*, and others.

they have just laid an egg. When Silkies get alarmed, they often have a laughing, almost monkey-like call.

Within those breed differences, you may begin to understand words. *This* call means, "I just laid an egg." *That* call means, "Mom is scattering treats." And *this* call means, "I have something delicious, but there's not enough to share." As the chickens forage, if you're close enough, you may hear the soft, companionable back-and-forth chatter that lets them know they are close enough to everyone to receive any important communications that will help keep them safe (and also that they will be within close range if their rooster locates something tasty they might be interested in eating). Dominant hens may issue warnings to others to wait their turn at the nest. A broody hen may growl or screech to tell others to stay away entirely.

Your chickens will even let you know when they have knocked their waterer or feeder over. You may begin to recognize the tone in their bagawking when they are complaining about something wrong in the coop, and you may learn to distinguish that from their "I just laid an egg" song.

When it comes to flock dynamic and culture, if you add or remove birds from your established flock, it can upset the pecking order and cause issues. After all, if Bessie is gone, which one of the hens below her in the pecking order will get her old spot on the roost at night? They will squabble for it. If there is a new bird, where will she sleep— will she try to claim someone else's spot? There will be trouble. In fact, your birds will have at it like a group of selfish middle schoolers.

The dynamic of your flock—the way they interact with one another—is based on their individual personalities.

Contrary to what one may hear from the industry, chickens are ... complex behaviorally, do quite well in learning, show a rich social organization, and have a diverse repertoire of calls. Anyone who has kept barnyard chickens recognizes their significant differences in personality.

—BERNARD ROLLIN, PhD, distinguished professor at Colorado State University

Each flock more or less develops its own culture, which means that a hen coming to your flock from another flock will probably be behaving gauchely, from the perspective of the established flock. She won't be doing what she's expected to—how rude!

For example, a "stay away" warning issued to a submissive hen in one flock might be an invitation for her to retreat to a far corner of the coop. In another flock—due to differences in the size of the coop, and the personalities involved— a similar warning might instead require a retreat of only a few steps, or it might require the offending hen to leave the coop entirely. Failure to adhere to the social contract is an insult of the highest order!

Your established flock will try to get the new member(s) to submit—to behave like civilized chickens, harrumph. In some cases where the coop/run space is at a premium, any new chickens could get pecked, seriously hurt, or even killed. That is, it's something that could happen unless you take the time to respect the pecking order and introduce new birds to the established flock properly. Taking the time to properly introduce your birds is essential, and you want to think about this *before* you order or hatch additional chickens. To read about how to safely introduce new birds to your established flock, turn to page 177.

The Case for Roosters

Roosters are wonderful to have in the flock for a number of reasons. Often, they have more showy, colorful plumage than hens; they are necessary for *fertile* eggs if you want to hatch your own chicks at home (although you don't need roosters for hens to lay eggs); they fulfill an important social role in flock hierarchy; and they can even help protect against predators by keeping a lookout and then warning your hens to run or take cover. There are also drawbacks to keeping roosters: They're loud, they can be aggressive, they're often prohibited in town, and they eat but don't lay. And keeping too many roosters can risk fatiguing and overbreeding hens, and fighting among your roosters. For the purposes of keeping a small flock of backyard layers, you normally don't want to plan for more than one rooster for every 10 or so hens. But as you've probably determined by now, sometimes not everything goes according to plan when it comes to pet chickens. If you end up with multiple roosters, we can offer a few tips.

5 Rules for Keeping Multiple Roosters in Your Flock

Roosters can be a challenge. They're great, don't get us wrong! They're often more spectacular-looking than hens—and can also be more friendly than hens are as chicks. They are usually filled with personality and charisma.

But they can also be territorial and protective of "their" hens. That's part of their charm, of course, but when they compete with one another too much, they can get hurt, or even hurt the hens! When you have multiple roosters in your flock, that protective instinct can get out of hand. Here are five simple rules to follow that will help keep the peace when you have more than one.

1. **Have plenty of hens for each rooster.** If you have a flock of only five to seven birds, you don't want two (or more) of them to be roosters. Generally, there should be 10 to 12 hens for each male in your flock. This will enable the roosters to have plenty of hens each, without worrying too much about competition from their rivals, and it will also be enough so that the hens don't get overbred. When there are too few hens for each rooster, a hen can be mated too often, resulting in broken feathers, bare backs and necks, or even injuries.

2. **Have plenty of space in your run.** When you have multiple roosters, there will be the occasional squabble, and, for the most part, that's okay. Those squabbles can get dangerous if there's not enough space, though. If your birds are too crowded—even

when there are plenty of hens—you may see serious problems. With multiple roosters you will need more than the bare minimum of space. You'll want to double or even triple the minimum space per bird for your flock. If you get too many roosters competing together in a confined space, testosterone-fueled aggression and territoriality can boil to a head. Remember, roosters don't have impulse control like humans do; someone could get hurt! If there is plenty of space, when one rooster becomes lax with his manners, generally the others will just lead "their" hens a respectable distance away, so they won't feel threatened by the boor and serious fights won't break out.

3. **If you have neither plenty of hens nor plenty of space, you can keep multiple roosters together by having *no* hens.** This is an arrangement you might have, for example, if you keep a flock of roosters for exhibition (rather than having a flock of hens for the purpose of laying). With no hens to compete for, multiple roosters often live together in relative peace.

4. **Raise them together in your flock.** Roosters who are raised together establish a pecking order among themselves as they are growing up. Because they have already established that order, there is less incentive to fight when they are older and more likely to hurt one another by sparring.

Alternatively, you can add new roosters to your flock relatively painlessly if they are raised by a hen in your flock, or if they are

introduced to your flock when they are young, before reaching sexual maturity. It will be difficult to maintain the peace if you add an adult rooster to a flock that already has roosters, because that new rooster will be regarded as an invader—not just by the other rooster(s), but also by your hens.

5. **Some roosters are too aggressive to get along with other roosters, no matter how ideal the conditions are.** There are some breeds that tend to produce very aggressive roosters that are prone to fighting, and other breeds with more genteel reputations. For example, game breeds (such as English Game, Modern Game, and American Game) often have highly aggressive roosters. Rhode Island Red roosters are notoriously aggressive, too. We've found that Easter Eggers and Ameraucanas don't always get along well with multiple roosters in the flock, either. That said, most backyard chicken breeds do fine in flocks with multiple roosters. Favorite breeds for roosters (and multiple roosters) include Salmon Faverolles, Plymouth Rocks, Marans, Orpingtons, Australorps, Silkies, and Brahmas.

Roosters attack with their spurs, the long, pointy, nail-like protrusions seen here.

Practical Housing Arrangements

Before you start worrying about *where* to locate a coop, you'll want to determine *how* you'll manage your flock, given your circumstances. Your conclusion will help suggest the best choices for your available space, coop, and run. So determine which flock-management style will be most convenient, and only then consider the details of your future coop and where you will put it.

In this section, we'll guide you through flock management styles, chicken coop features, housing and caring for baby chicks, and how to be predator-savvy.

FLOCK-MANAGEMENT STYLES

There are four very workable ways to keep a small flock: free ranging, confined ranging, part-time ranging, or full-time confinement. Let's discuss the pros and cons of each.

FREE RANGING

When a flock roams freely and forages in an open field or large yard, they are free ranging. Do you live far enough away from neighbors and traffic so that your chickens can rely on their natural instincts to keep them safe? If so, then you might consider free ranging your flock. In some ways, free ranging is the simplest way to manage your chickens.

However, true free ranging would be problematic for most people, since—no matter how many times you may warn them—your chickens won't recognize property lines. They won't know to stay within the boundaries of your land, and they certainly won't understand that they shouldn't roost on your neighbor's porch or find bugs in her yard.

And consider this: Unless you fence them out of areas, you'll have to be okay with the occasional poop on your own walkway or porch and scratching in your own garden. So, even if true free ranging is an option for you, will it be something you want?

In addition to the fact that free ranging requires a good deal of space and no nearby neighbors, there's also risk involved. When you free range, you risk losing the occasional bird to predators. For that reason, even when free ranging, you'll want to securely shut the coop door at night, when most predators are roaming and the chickens are asleep and at their most vulnerable.

You'll also need to open your coop again in the morning to let them out, unless you have an automatic door.

The "work" required with free ranging—opening and shutting the coop—is much less than what keeping a dog entails, with daily walks and frequent bathroom trips. Even so, consider that you'll need to be there *every evening* to close the coop. Seriously.

Every. Evening.

Depending on your latitude, during the winter you may have to wait until 6:00 p.m. or so if you want to leave the house in the evening, because you'll need to secure your flock before you go. In the summer, you might have to wait until 9:00 p.m. to leave. Or you may choose to leave the house early and return before or shortly after sunset, so you can shut the coop door and keep your pet chickens safe from raccoons, foxes, and other dangers. This responsibility can cramp your social life, making it difficult to spend the night elsewhere, because there is only a certain amount of time you can be away from home. The result could be late arrivals at dinner parties or early departures. You may have to travel separately from other family members so someone can stay behind (or leave early) to secure the coop. But you could get

Choosing Free Ranging

Free ranging is the easiest and least expensive flock management style, but it's not for everyone. Consider the pros and cons carefully *before* choosing your coop.

Pros

- Inexpensive; lower initial investment needed when you don't have to purchase or build an enclosed run.

- Reduced feed bills, especially in spring and summer, because the chickens will be supplementing their diet with what they can forage while out ranging.

- Reduced need to clean the coop since most time will be spent outdoors; no run to clean or maintain.

- More space means fewer pecking order issues.

- Can get breeds that don't tolerate confinement well.

- Chicken eggs will be more nutritious with more access to pasture.

Cons

- Large space requirements.

- Greater dangers from predators.

- Still some commitment in time unless you acquire an automatic coop door.

- The chickens will be more exposed to illnesses and infestations from wild birds.

- They may find places to lay that you don't know about, so you may "lose" some eggs to hidden nests.

an automatic chicken door, so you won't have to worry about manually opening and shutting the coop. It can be a fabulous work-saver.

An automatic coop door is an expense, although it will be cheaper in most cases than the cost of fencing a large, secure run. Depending on the model, an automatic coop door may be set on a timer or have a light sensor so that it automatically shuts at night and automatically opens in the morning. But, whether you get an automated door or not, there is a certain amount of commitment that comes with choosing to free range. If you decide to manage your flock this way, be sure that you're okay with that level of commitment, whether it's a commitment of time (shutting and opening the coop), or whether it's a commitment of money (buying an automated door).

CONFINED RANGING

Confined ranging is one of the most common ways to manage a small flock in an urban or suburban setting. In this system, your birds have access to a limited yard or run. In other words, they range in an enclosed area like a fenced backyard or a separate fenced run, rather than having free access to wherever their birdy brains may take them. With fences or a secure run, your flock can't wander willy-nilly onto a neighbor's property or out into the road. Even in rural settings, if your neighbors are nearby enough that your chickens may trespass, having a fenced yard or run may be the way to go.

Even with small, suburban-size yards, the confined ranging system can work well. Presuming the run is large enough that your flock has access to green pasture, confined ranging often works out to be much the same thing as fenceless ranging, so far as the birds are concerned. The birds themselves don't get to define their territory like they do without fences, but they do have a large outdoor area that they can explore freely during the day. Another benefit is that, as with true free ranging, your feed bill will be reduced because your girls will have access to lots of forage.

Here's the thing about confined ranging, though: If their area is too small, your hens will forage the dickens out of it until there is nothing left but bare earth. They will still have bugs to eat (plus whatever treats you can offer), but if you don't provide them with enough grassy yard, their eggs will lose some nutritional benefits.

That's okay, mind you. Your eggs will still be super fresh, and there's a great peace of mind in knowing that your chickens are being treated humanely. Plus, chickens are fabulous pets, and would be even if they laid zero eggs. Still, when you're looking at all the implications of choosing one management style over another, the effect no pasture will have on your eggs is something to consider. So that means you'll need to fence a large space—and that means expense.

Or does it? There is a way to manage confined ranging inexpensively with little space, while providing fresh pasture: the chicken tractor. The term may make you laugh—chickens driving tractors?!—but the system is ingenious, and doesn't involve getting a John Deere or teaching your chickens to manage heavy farm equipment.

Chicken tractors are mobile coops that have small, attached runs. The operative word here is *mobile*: Tractor-style coops are simply lifted, and scooted or rolled (depending on the design), every day or two so that the run area is always located over fresh pasture. This means that no grassy area gets overworked or overpooped-on. Even though a chicken tractor's run can sometimes be small, the chickens have access to new range every day, and they don't suffer from boredom or lack of nutrition. The chickens can spend a day or two with the tractor in one spot—they scratch and forage and fertilize—then the tractor is moved to the next spot, where they can enjoy the same thing again.

This is a truly ingenious method of flock management, especially for urban and suburban chicken keepers with limited space and/or close neighbors. The birds are not only always safely confined, but they are also always on fresh range so they will lay nutritionally superior eggs. Remember, without access to a grassy yard or pasture, their eggs won't be nutritionally different from grocery store eggs. Plus, with proper planning, you can coordinate being able to fit the tractor between your garden rows on occasion, so your chickens can do the weeding for you. How cool is that?!

Tractors are best for small flocks of a few birds (up to 10 or so), since larger tractors can be too heavy for backyard chicken keepers to move. For example, the Eglu Cube is made of lightweight plastic and holds up to 10 chickens in a 2- by 3-foot area plus a 6- or 9-foot run, though we would recommend limiting the number to six unless you've chosen a small breed. This size will accommodate the vast majority of backyard flocks, though. If you really want to keep more than 10 or so birds in a tractor, you may have to get multiple tractors so each group can be moved, or else you'll need some equipment to help move the coops. Some forward-thinking

Allowing your chickens to free range is a gift for both of you. (The healthy, beautiful eggs! The overjoyed birds!) If that's not an option for you, confined ranging or a movable, tractor-style coop are the next-best choices.

commercial farmers raise their birds in very large chicken tractors that are moved with heavy farm equipment.

There are two catches, though. One is that this method is not ideal for some breeds, since some don't do especially well in close confinement. Just like you wouldn't want to get a large, high-energy dog if you lived in a studio apartment with no yard, you wouldn't want to keep flighty, high-energy chicken breeds with a need for extensive range, such as Hamburgs, in a small, tractor-style coop. That said, most of the popular backyard breeds do quite well in trac-

tors. So unless you're dying to get a flock of aggressive birds or skittish, nervous flyers, it won't be a problem in most situations.

The second catch is that for a tractor to be a good flock-management choice for your situation, you'll need a flat-ish yard. This doesn't mean your yard has to be absolutely level, but it does mean that the landscape needs to be such that your tractor-style coop won't be leaning precariously on the side of a hill all the time. Your yard also needs to be flat enough so that the bottom of the coop effectively meets the ground, excluding predators and keeping your birds safe.

Choosing Permanently Sited Setups

Permanently sited coops are a great choice in any chicken management system.

Pros

- Keeps your chickens out of your neighbors' yards!

- Foraging reduces feed bills and alleviates boredom and aggression.

- If your yard is already fenced, it can be inexpensive.

- Permanently sited coops are less expensive than tractor-style coops.

- They are less physically taxing than a coop you'll have to move.

- You won't need to worry about nocturnal predators digging under your permanently sited run, since your chickens will be safely enclosed in the coop each night.

Cons

- Even when your yard is already fenced, the chickens may not stay within it if you get breeds or varieties that need a lot of range or like to fly; so you'll need either a tall fence or heavy-breed chickens!

- Building a new run or fully enclosing a large run, can be expensive, because you'll need to fence a relatively large area to maintain grassy forage.

- Unless your run is completely secure (including enclosing the top), you will still have to commit to opening and shutting the coop door (or getting an automatic one), since raccoons and other chicken predators can simply climb over a fence to get at your chickens.

PART-TIME RANGING

The third way to manage your hens is part-time ranging. This is a great way to operate in some situations. For instance, if you don't have the financial resources to build a secure run that is large enough to avoid overgrazing, then you might choose to build a small, secure run and let your birds out to range occasionally. This way, you'll control when they are let out to forage. This also may be the choice you need to make if you can't—or don't want to—commit to being home at sunset every night to shut the coop door and keep your girls secure from predators.

With part-time ranging, you may have a tractor or a permanently sited coop with a small attached run. (Even those with chicken tractors may choose to let their birds out to range more widely, socialize with their humans, and enjoy the extra space from time to time.) Many people only let their birds out in the evenings, when they will retire on their own when it's time. The saying "chickens come home to roost" is really true. Once your chickens learn where "home" is (which you accomplish by keeping them enclosed in their coop for 2 or 3 days when you

Choosing Tractor-Style Coops

Tractor-style coops are especially popular in small, suburban yards.

Pros

- Keeps your chickens out of your neighbors' yards!

- Keeps your chickens excluded from your own vegetable gardens, landscaped beds, patios, sidewalks, and other places you don't want them.

- There is no time commitment involved to open or shut the tractor-coop door in a fully enclosed, secure run.

- A chicken tractor and run is small and relatively inexpensive to build. While it's usually more expensive than simple free ranging since an enclosure must be built, it's still one of the most economical ways to manage your flock.

Cons

- Requires relatively flat land.

- Be prepared to move the coop every day or two.

- Be sure to choose breeds that can tolerate close confinement.

- Tractor-style coops tend to be costlier than their conventional counterparts.

- Movable coops are not fully predator-proof unless the run floor is enclosed with a sturdy, welded wire mesh. The wire floor hinders a bird's ability to access certain goodies (like worms!) and to dust bathe. You will have to provide an artificial dust bath in this case.

- Unless you have the heavy equipment needed to move a large coop, you'll have to choose a smaller coop (holding fewer birds) that can be moved by hand.

first get them), they will return home at dusk every evening. It's one of the biggest surprises for newcomers to the hobby.

You may decide it's necessary to supervise ranging time if your neighborhood has many stray dogs, or if there is a large risk of danger from other predators. Or if you're not home to supervise during the day, your family may decide to let the hens out only in the evenings or on weekends when someone is around, in case there is any trouble. Part-time ranging is a popular choice.

If the run is very small, it is recommended that you build a roof over it, so that you can keep dry, clean bedding inside. After all, you don't want to confine your chickens to a dirt floor that gets overly muddy—or worse, you don't want to put bedding in your run to keep them out of the mud and then have it get moldy or begin to rot! For those reasons, a small run—even when your flock is ranging part-time—will require more time, effort, and expense to keep clean, in contrast to a large one where droppings are widely scattered and will simply fertilize the lawn and dissipate.

Choosing Part-Time Ranging

Part-time ranging is the preferred choice for most backyard chickenistas: It's relatively low cost, and it strikes a balance between protection from predators and the good nutrition offered by free ranging.

Pros

- Usually less expensive than simple confined ranging, because the area being secured is small.
- Your birds will remain safe from predators when they are inside their secure run and, for the most part, if you provide good supervision during times of ranging.
- Superior to full-time confinement when it comes to increasing nutrition and reducing feed costs and flock boredom or aggression.
- In daily commitment, it is superior to free ranging and confined ranging (without an automatic coop door), since you can keep them in their secure run when needed, and you won't have to be at home to open and shut the coop door.

Cons

- Your birds will have less access to green forage than they would if they were ranging all the time, so feed costs will be higher than with free or confined ranging.
- Eggs will be less nutritious than the eggs of full-time pasture-raised hens.
- While enclosed, your birds may be more prone to boredom or aggression than birds that range full time, since there will be much less to occupy them when foraging in a small run without grass.

That can be an advantage, though: Chicken poop makes great fertilizer—that is, presuming it's composted and not all concentrated in one spot! Chicken manure is very "hot," meaning that it has so much nitrogen, it can easily burn plants unless it's been composted and aged or widely distributed over a large area. A small run doesn't get defoliated and turn to dirt just because the chickens graze it, although that certainly plays a part. It's also a matter of all those droppings being left in one small area and causing the plants inside to get overfertilized and die.

FULL-TIME CONFINEMENT

Some urban and suburban keepers don't range their birds at all, but instead confine them full-time. Full-time confinement means a flock has access to the outdoors, but not to green pasture—not even on a part-time basis. Even when backyard chicken keepers confine their pet chickens, the flock is usually still quite pampered.

The birds may not have pasture to forage on, but they probably get scraps of fruits and veggies as treats to supplement their diet. Their owners may also buy special treats like mealworms, scratch, and sunflower seeds. Some will even cook for their birds. See Chapter 15 for recipes you can make for your chickens.

Even confined space provided for pet backyard chickens can be plenty, presuming you observe per-bird space recommendations of allowing 10 square feet per bird. The advantage of not ranging at all is that if your coop and run are secure, there is absolutely no danger of predators. The disadvantage, though, is that there will be a lot more cleaning involved and birds will be more prone to boredom and behavior issues like pecking and egg eating. Illnesses and infestations can also spread more quickly through the confined flock. Plus, feeding them may be a bit more expensive, since the birds will have no ability to supplement their diet by foraging.

Factory Farm "Free Range"

Factory farm–style "free ranging" is not actually free ranging. It would be classified as full-time confinement. In commercial farms, free-range eggs come from hens that should be free to go in and out of a barn or protected area, roaming in an outdoor pen for part of the day. However, the size of the space or the level of access to that outdoor space is not regulated. The outside area may be as small as 10 feet by 10 feet for 20,000 hens; is not required to have any dirt or grass; and is usually restricted by a small door, resulting in no meaningful outdoor access for the vast majority of hens raised in this manner.

So just to clarify, full-time confinement by caring pet chicken owners does not at all equate with commercial confinement.

WHICH STYLE IS BEST FOR YOU?

So, which of the four flock-management styles should you choose for your family? Any of them will work. If you're not sure which style will be best for you, take a look at this axis, going from least to most range restriction:

Free ranging → Confined ranging →
Part-time ranging → Full-time confinement

Ranging has a big influence on your birds' lives, and your choice of management style will affect many things about your flock. Every method here is valid, but you'll want to go in with your eyes open to both the pros and cons of the methods you're considering.

On the left side of the spectrum illustrated above—all other things being equal—the more range there is, the less likely you are to have behavior problems stemming from the boredom of confinement. The eggs will be more nutritious when your birds have access to pasture since foraging provides more of the vitamins needed for egg-making. Feed costs will be somewhat less with more range (although in the case of a small backyard flock, your savings will not amount to a great deal). Bedding costs will be reduced, too: Cleaning will be required less frequently if your birds spend most of their time outside fertilizing the yard rather than in the coop or a small enclosed run, accumulating manure that will need to be composted or otherwise disposed of. Plus, with more space, it will be easier to introduce new birds to your flock, if needed.

On the right side of the spectrum, your birds

Choosing Full-Time Confinement

Even though full-time confinement is the costliest management style and has some unique challenges, by providing large, predator-proof runs, your flock will be much happier and healthier than they would be in any factory farm setting.

Pros

- Full-time confinement is the most secure of all management methods since your birds will be in their protected area at all times.

- You won't have to be at home to open and shut the coop door; in fact, you won't even have the occasional time commitment of opening and shutting the door on evenings or weekends.

- Your birds will all be confined away from your gardens and landscaping, as well.

Cons

- Full-time confinement creates the most issues with flock boredom and aggression.

- Most expensive in terms of feed, since the birds won't be supplementing their diet by ranging.

- Unless you provide your own supplements, your flock's eggs will not be as nutritious as they would be if the hens had access to pasture.

Free-range birds get the choicest tidbits and "fertilize" the yard as they go.

will be more secure as confinement increases. And while confinement may require more maintenance and expense, it also means you will use less space for your hobby. With some confinement, your garden beds won't present such attractive targets for wayward foragers and dust bathers. When a hen or two decide to hide their eggs in the yard, the eggs will be easier to find because there will be fewer places to hide them. Further, with very secure systems, you won't have to worry about being home at certain times to open or shut your coop door. And, finally, although more cleaning will be involved with more confinement, this also means you will have access to more material for composting. If you garden, it's like hitting the jackpot.

If there is more than one system that will work in your circumstances but you're torn as to which to choose, weigh what benefits are most important to you. For instance, if the increased nutrition of backyard eggs is your chief aim, then you may want to lean toward more ranging time than less. If you have a green thumb, you may prefer to keep your birds confined away from your gardens and take advantage of the increased compost.

Whichever system you choose, be sure to go in with your eyes open: You will be accepting not only the benefits, but also the risks associated with your choice.

Egg-cetera!

fresh eggs

When there is a large ratio of females to males in a flock, sometimes one female will take on a male social role and may even begin to crow. She may also cease laying and can start to develop long rooster-like spurs from her spur buds.

Lissa's Scratchings

It's important to be ready to respond to threats when your flock is free ranging. A few years ago, we had some issues with a persistent (and quite fearless) fox harassing our free-ranging flock. The fox was so quick and daring that he even made daytime raids, trying to carry off birds in broad daylight—it was terrible!

One day about midmorning, I was inside on the computer, plugging away for My Pet Chicken. It was before I'd even changed out of my nightgown. (One of the weird things about working from home is that you may not get to shower and dress until your workday is over!)

From the yard came a warning cry from my rooster, Gautier. I recognized that cry: He was letting everyone know there was a predator coming—a predator on the ground, rather than from the sky! I rose immediately and headed for the door, but before I could even slip on a pair of sandals, Gautier's warning was followed by several screeches, accompanied by the sound of many flapping wings and general chaos! In a paroxysm of fear, I forsook shoes and bolted out the door, barefoot, running toward the sound as fast as I could to try to save whoever was hurt or in danger.

I could see Gautier at the edge of the yard where the mountainside dropped off past an old barbed wire fence. He was a mass of indignation, his hackles raised to such a degree that he appeared to be sporting a rather elaborate Victorian ruff. I've never regarded ruffs as anything other than outmoded fashion affectations, but on Gautier the effect was fierce. He meant business.

So did I.

I'm not athletic, but I somehow bounded over the barbed wire fence in front of him without breaking stride. I might have been flying. Or perhaps I teleported. I shudder now to think what could have happened had my nightgown—or, for that matter, my leg—caught on the razor wire and sent me vaulting down the mountainside face first. In fact, I've never been able to explain satisfactorily, even to myself, how I got to the other side of the fence without injury.

However it happened, though, I did it. Once past, I sped down the rocky, multiflora-rosed mountainside in my bare feet like I was a professional stunt person making the descent by sliding on an invisible zip line. In my head, I heard Samuel L. Jackson's voice thundering that biblical quote from *Pulp Fiction*: "And I will strike down upon thee with great vengeance and furious anger those who attempt to poison and destroy my brothers!" Beside me Gautier flew like a black and gold bullet, his every feather perturbed, making him look about three times larger than he had any right to be.

Finally, I saw below us what I was after: the predator that had attacked my flock. It was a fox. In front of the creature, and mostly out of sight behind a tree, huddled a disordered mass of feathers, completely unmoving. As soon as the fox realized he'd been spotted, he melted away into the forest, leaving the heap of feathers behind. Gautier tailed him briefly, but the fox was considerably faster than we were, even given my possibly magical transportation over the barbed wire fence. Gautier emitted a final harrumphing rebuke, and turned contemptuously to scratch the earth up in the fleeing fox's general direction before returning to my side, near the victim.

I had come to a halt in front of the feather pile, my heart thumping. It was my hen Marissa; I recognized the unusual plumage of my half Rhode Island Red, half Faverolles bird. Now that the fox had been driven off, I almost didn't want to see how badly she had been hurt. I didn't want her to be dead or, worse, suffering from some horrific injury the fox had inflicted. I also dreaded the prospect of actually having to put her out of her misery if her wounds were too severe. I hadn't hesitated when I was chasing the fox, but now I was momentarily paralyzed.

But only momentarily. If there was anything I could do, I needed to know, and quickly. And if she wasn't under there, I needed to find her. Finally, I knelt down in front of the feather pile. I still couldn't quite tell if this heap was the body, or if the body was elsewhere, and this was just a pile of plucked plumage. Gautier sidled up beside me as I put a trembling hand out toward her feathers to see if she was beneath.

As soon as I laid my hand down on the mass, I felt chicken beneath, and the feathers arranged themselves into a recognizable form. Her little chicken head popped up and looked around, blinking and disoriented. Gautier and I both startled a little. But she was alive! She made no noise, though. Was she hurt? I carefully reached beneath her body to lift her up. I didn't want to exacerbate any injuries or cause her any more pain if something was broken. She was eerily quiet.

To my utter amazement, I didn't observe any injuries on her. There was no blood that I could see—at least nothing I could see in the shade—so I tucked her under my arm and carefully carried her up the slope where I could get a better look at her in the yard out of the shadow of the mountain. After all, she might have puncture wounds from the teeth that dragged her down here, or broken bones.

She remained silent during the ascent, perhaps in shock. I, on the other hand, was starting to realize that my legs had been shredded in my flight down the mountain. I had missed the barbed wire, yes, but I seemed to have torn directly through every wild blackberry bramble and thorny rose bush on the hillside. Plus, I had surely stepped on every stone and branch with my tender feet. I hadn't noticed during my flight down the mountain, but I could now see that my legs and feet were bloody and bruised.

And, as it turned out, mine were the only actual injuries in the whole adventure, and they were relatively minor considering the complete and utter disregard I had for anything other than getting to Marissa in time.

It was worth it, too. She would most likely have died otherwise, and perhaps Gautier along with her.

So, what's the moral of this interlude? Perhaps Aesop would suggest something like, "A rooster who cries 'fox!' doesn't lie." Or maybe, "Look before you leap, unless you're on a rescue mission." However, what I think it illustrates most importantly is that "it's too late to whet the sword when the trumpet sounds." In other words, you should do your best to be prepared for emergency situations before they arise. While you can't avoid every hazard, you should *be aware* of the particular risks you chose to accept when deciding on your management style, and you should know what actions to take when necessary.

For myself, I must accept that there will always be a danger to my free-ranging birds from predators. Not everyone wants to suffer that risk. It's okay to prefer a small, enclosed run safe from predators, and instead accept the increased risk of flock boredom or aggression—provided you are prepared to handle the particular risks accompanying your choice.

CHAPTER 6

The BROODER and COOP

With some planning ahead, taking care of your chickens will be a piece of cake! If you have a busy lifestyle, it's especially important to invest in features like electricity and a light in your coop, along with an automatic door and automatic-style waterers and feeders.

Once you have chosen your management style, you can address the actual siting of your chicken coop. If you have decided on a mobile tractor-style coop, you'll be able to maneuver it into the most convenient locations, as required. But for permanently sited coops, consider these questions.

• How far do you want to walk to gather eggs, taking into account possible inclement weather?

• Where will you be refilling water and feed—and how far do you want to carry a filled waterer or feeder? (Or, do you plan on using a waterer that hooks up to your garden hose?)

• Are you a light sleeper? How close do you want the coop and/or run to be located to bedroom windows?

• How close do you want your coop to be located to your compost pile or to the garden?

• Are there setbacks required in your area (meaning, do you have to locate your coop a certain distance from property lines)?

• If summers are very hot in your area, what sort of shade will your chickens have during the day?

• If winters are very cold in your area, what sort of sun will they have?

• Does the area have the potential of getting muddy in wet weather? Mud gets on chickens' feet and feathers, and dirties eggs. Mud that stays around is even worse, turning the run into a swamp.

• Is the ground level enough—or can you make

it level enough—to install your secure coop and run?

• If you plan on watching your chickens for entertainment, can you see the chickens' area from where you'll want to be sitting (on the porch, on a patio, in an arbor)?

• Are there any hazards in the area such as a pool or pond from which you will have to exclude the chickens?

• Is there anything toxic or dangerous growing in the area that you will have to remove? For instance, while chickens aren't generally allergic to poison ivy, if you intend to pick them up and pet them, you may not want them getting poison ivy oil all over their feathers and transmitting it to you!

• If you think you'll want to add a heated waterer, an automatic chicken door, or other electronic equipment to your coop, will it be close enough to a power source? Or, if you choose to power your coop with solar, will there be a good site for that near your coop?

• What is a reasonable distance to locate your coop away from roads, so headlights and street noises do not disturb your birds at night while they're trying to sleep? On very quiet roads it's not an issue, but where the speed limit is 30 mph or more, keep coops at least 50 feet away from the street.

• What is a reasonable distance to locate your coop away from your dog(s) or your neighbors' dogs, so your chickens do not get intimidated or scared?

The list can continue and get more and more specific, but you get the idea. Think everything through—determine how *you* would like to site your coop to make your hobby easy and enjoyable for your family. And, remember, moving a permanent coop can be near impossible, so you want to get it right the first time!

IMPORTANT—AND FUN— COOP FEATURES

Which features you'll want in your coop will be a matter of personal preference, but it can help to know the pros and cons of each. Here is a rundown of coop features.

Automatic door. You will probably love having an automatic coop door. However, it's one of the more expensive "extras" you can get for your coop. And if your run is already fully secure (including from above and below), an automatic door is not really needed. This is most helpful in a free-range or confined-range situation,

Egg-cetera!

fresh eggs

Before a hen begins to lay, her comb will get larger and redder, and she may begin to squat submissively when you reach down to pet her.

particularly if you can't always be home to close your chickens in at dusk.

Green roof. Some chicken coops have what is called a green roof, meaning the roof holds soil and is planted with flowers, herbs, or other plants. This feature provides good insulation and can also be beautiful. But if you're not a gardener, you may not want the extra work associated with keeping that bed watered and healthy. Plus, if your chickens have access to the top of the coop, they may be tempted to use it as a dust bath—not nearly as pretty. Also, you'll have to be particular about what you plant so it doesn't get eaten. Herbs might be a good bet, though chickens' plant preferences vary, so you'll have to experiment.

Roosts. It's nice to have built-in roosts. In most cases you can add your own, though, if your coop doesn't come with any. Look for roosts that are wide and flat and made of wood. Round roosts mean that your chickens will have to grip, and their toes can stick out from beneath their feathers in chilly weather, making them cold and uncomfortable. Metal also seems like a great, durable idea—but icy metal roosts can contribute to frostbitten toes in winter. Make sure your roosts are higher than your nest boxes, or your girls may try to sleep in the nests, making them dirty. Nests are for egg laying, not sleeping.

Automatic feeders and waterers. Some waterers and feeders have to be refilled daily, whereas the automatic type may only require refilling weekly (or never!). These rely on gravity to deliver more feed or water to your chickens as they consume it. Some bowl-style fully

This small coop features two roosts, a nesting area, and a slide-open ventilation slot.

automatic waterers and nipple waterer systems actually hook up to your hose so you never have to refill them, so long as the air temperatures are above freezing. (You may need to switch to a heated waterer during winter months, so consider your proximity to electricity.)

Nest boxes. Not all coops will have built-in nest boxes, and that's okay. They're easy to buy and add, in most cases, but consider the options carefully. You'll need about one nest for every four birds or so. If you have only three or four birds, it can help to have an extra nest. If you have an extra, when someone goes broody and won't leave the nest, you'll still have an accessible nest that your other girls can use. Be forewarned that even if you have a nest box for everyone, there is sometimes just one favorite nest that they all prefer, while the others go empty.

Roll-away nest boxes. These are a great solution if you're having trouble with egg eaters. When the egg is laid, it rolls back away from the hen and into a collection area she can't get to.

Plastic nest boxes. Plastic nest boxes are easy to clean and sanitize, and, unlike wooden nest boxes, they don't have as many good places for mites or bugs to hide. However, depending on the design, they may take up more space than traditional wooden nests.

Wooden nests. Wood is the classic material from which to make nests. They can be something of a pain to clean and sanitize, but you'll be doing this only once a year, unless there's been a cracked or broken egg (or another unexpected mess). Consider painting the boxes (use milk paint or another nontoxic paint) or lining them with linoleum to make cleanup easier.

Built-in nests. Most coops have their nests built right in; it's a basic feature. If your coop doesn't come with nests, though, it's not a big deal: They're easy to purchase or build! Even if you're not handy, nest boxes are simple. A typical nest box is 12 inches cubed, with a lip at the bottom to prevent bedding material from falling out. Still, you'll want to add this cost into your calculations when you're deciding which coop to purchase or build. Cost out your project at your local hardware store—prices vary in different locations.

Bucket nests. If your coop doesn't have built-in nests, you might decide to use cheap bucket nests. You can lay a 5-gallon plastic bucket on its side and pad it with nesting material. Add a brick or a stone on either side of the bucket to prevent it from rolling, and you're set. These have the advantage of being easy to clean like commercial plastic nests. When nests are directly at ground level, though, it can mean that it's hard for a girl to get any privacy to lay. Keep in mind, too, that some hens won't appreciate how bright it is inside a white bucket, as they prefer to lay in dark places, so you may want to choose a black or dark-colored bucket.

Crate nests/other makeshift nests. Your hens really don't need anything special in which to lay their eggs. No matter how fancy your nest boxes are, you may find some of your hens prefer to lay outside beneath a shrub, or in a corner of the coop, or even in a flowerpot on your porch. One of Lissa's girls inexplicably went through a period of wanting to lay eggs on the flat, hard, bright white top of a chest freezer. What gives, Prissy? So, many people simply use what they have at home to make nest boxes: cat litter boxes, old drawers, plastic storage tubs, or other makeshift containers. Lissa's Silkies preferred a tangerine crate laid on the floor in the corner of the coop, and ignored her nice big plastic nests.

Outside egg door. Coops with built-in nests often also have an outside egg door, meaning you don't have to go into the coop to gather eggs. Nice feature! However, if you have a coop with an outside egg door, you'll want to be sure the door is secure against predators like raccoons. Raccoons are pretty wily and can often open simple latches and locks.

Add-on runs. Some coops have a small built-in run, which is nice for days when your flock may have to stay inside. Some also have "additions" that may be purchased to add more space to your built-in run. This can be a convenient way to expand your space slowly as you expand your flock.

Windows. Having some light inside your coop is generally a good thing—but a lot of light is not always needed, or even desired. Unless you live in an area with significant severe weather, your chickens will be outside in their run (or free ranging) most of the time, rather than inside the coop. They will be more active when it's bright out. A dim (but not dark) coop will be more conducive to egg laying and less conducive to egg eating and aggressive picking.

Windows that open. It's nice to be able to add extra ventilation on warm days. If you live in a hot area, having opening windows on your coop can be especially nice. Windows can provide ventilation if your coop doesn't have ventilation holes or gaps. Make sure they're secure when they're open, though, or predators could get in. A regular window screen is not secure; you'll want to cover the windows with something like fine-mesh ($\frac{1}{2}$- or $\frac{1}{4}$-inch) hardware cloth, which is a welded wire product.

Ventilation holes. If your coop doesn't have windows that open, you'll want to make sure there is some ventilation built in, to keep the air fresh and clean and let out moisture. Chickens are especially prone to respiratory illness, so ventilation is a must. As always, make sure they're secure against predators.

Walk-in coop. A walk-in coop can be nice. Carry in a chair or a bucket and watch your birds socialize. It's simpler to clean the coop if you can get to everything easily. However, a walk-in coop

Most coops come with built-in nest boxes like these, with a dedicated door for easy egg collection.

is usually an expense. Chickens don't need ceilings that high. Plus, if the coop is very roomy, their body heat doesn't make it quite as warm in the winter as it would in a smaller area. On the other hand, if you have a large walk-in coop, you will have space for a brooder or hospital area right inside the coop, which will make many things, such as introducing new birds to your flock or rehabbing an injured bird, simpler down the line.

Lift-up roof. Sounds great! But make sure the roof is low enough that you can reach everything you need to. If it opens directly into the coop—and if that is the only access—you'll want to be able to reach roosts for cleaning, feeders for filling, waterers, nests, and/or the floor to gather eggs. If your coop with a lift-up roof is too tall, you may not be able to perform basic maintenance without a struggle, and that's something you want to avoid. Some aren't designed so well, and require standing on a ladder to reach the floor of the coop from the outside. You'll also want to make sure the roof is not so heavy that you have difficulty lifting or holding it—but not so light that it will blow away in the first wind. And, of course, it'll need to latch securely against predators.

Brooder or hospital area. If your coop is large enough, you may have space to add a separate area right inside the coop where you can keep a hen who is recovering from an injury. (If your hen is ill with something contagious, you want to keep her away from the other birds.) Or you may use it to keep a broody and her babies safe. If you have space to keep new birds inside the coop—separated by wire—that will make final introductions *so* much simpler.

ELECTRICITY

Having electricity in your coop can be a big help. Even if you're sure you won't use an automatic door, a light, or a water heater to prevent freezing, a year or two into the hobby, you could feel differently, and it may be less expensive to have electricity installed upfront rather than to have your coop retrofitted down the line. Chicken coops are dusty, however, and electricity can cause fires, so be sure to get help from a licensed electrician, and be *especially* cautious with high-wattage heat lamps.

Adding Light to Your Coop

Many people add light to their coops in the winter to help boost winter laying, but this is not always such a good idea. Laying decreases in winter because the short days don't stimulate the hen to lay eggs, so it makes sense on the surface to think that adding light could solve this problem.

Here's how it works: Light cues tell your chicken's body whether or not to release an ovum (the yolk plus the hen's genetic material) from her ovaries. It doesn't have to do with cooler winter temperatures, as some think—at least not primarily. Even when the winter weather is warm, if there are not enough hours of light in the day, your chickens will lay less frequently. It's true that very cold weather can also contribute to a drop in laying, because more of your chickens' bodily resources have to go into keeping them warm rather than producing eggs. And it doesn't help either that short days mean there's less time to eat, so they consume fewer calories. However,

Adding a light to your coop can be helpful, whether or not you use it to increase egg production in winter. After all, it will be dusk when you close your birds in for the night!

the primary reason for a winter slowdown is the light itself. Your chickens have what is colloquially referred to as the third eye—the pineal gland on the surface of their brains that perceives light. In some other animals, specifically some reptiles and amphibians, the pineal or parietal eye is well developed—it's an actual eye that can be seen on the surface of the skin of the head. In chickens, the light is thought to penetrate the skull and reach the pineal gland that way. It's not *really* a third eye, not for chickens. But even blind chickens respond to changing day-length—strange but true. If you'd like to read more about this, visit bluetoad.com.

So if that's the case, why wouldn't you want to add light for more eggs? There are two good reasons. First, light cues have other effects on your hens. For instance, shortened day-lengths are also a cue for your chickens to begin molting. So one problem with keeping the daylight constant is, if you add light to the coop, it could cause your birds to molt late, in the dead of winter, when it is cold and they need their feathers the most. Chickens will molt annually, regardless of the light situation; however, it is normally the change in daylight hours that triggers it (not temperature changes). If your chickens don't have that trigger from fading hours of light, they may hold on for several months before finally molting at a time when it is really too cold to be without feathers. Yikes!

Second, if their brains are telling them to lay eggs when it's very cold outside, the resources going into the production of those eggs may well be resources they could otherwise be using to keep warm. The constant peak production can add undue stress to their systems, when it would be better for them to have a natural rest to recharge their batteries.

Even knowing all that, many people still decide to add light to the coop. In a poll of our customers, with no alternative sources of *real* eggs, about 21 percent will go without rather than be forced to eat store-bought. But if you do feel you must add light to your coop in the winter to increase production, at least do it the right way: Only add light in the morning, and only after your chickens have had their annual molt.

THE BROODER

The coop is where your chickens will live as adults, but if you start with baby chicks, you'll need a *brooder*, not a coop.

You'll recall the term *broody* from Chapter 3 in our Breed Selector Guide. When *a particular hen* is broody, this means she wants to hatch her eggs. A *breed* that's described as broody has hens that often, individually, go broody. A *brooder* is the housing that your baby chicks need to keep them safe and healthy, and it's also a term used for a mother hen who is brooding chicks.

When you purchase day-old baby chicks, you can't simply put them directly in the coop they'll live in when they're older; baby chicks need special care, especially warmth, since they can't maintain their body temperature well for the first few weeks. If you don't have a brooder (hen), then you'll need a brooder (housing).

Got it?

Brooder Boxes and Other Things Chicks Need

Let's talk first about a brooder (which, in this context, simply means housing).

First you need a basic enclosure—somewhere the chicks will live. It can be created from all sorts of things; it doesn't have to be anything fancy. Some people buy brooder kits that contain everything needed. Certain kits simply have a roll of thin cardboard you can use to cre-ate temporary brooder walls. My Pet Chicken has a nifty brooder kit that contains sturdy, double-walled cardboard panels that lock together to create an enclosure. You can also use a large plastic storage tub covered with netting or wire—or you can use a kiddie pool or even a large cardboard box.

There are some commercial brooder boxes, too, that—for a few hundred dollars—come complete with an enclosure, a heat source, and feeders and waterers all built in. But be forewarned: Some of these commercial brooders will claim to house 50 chicks (or more), which sounds great. But they'll only house that many if the chicks are very densely packed together. It's not nearly enough room, and it will create all sorts of aggression and other bad behaviors in your chicks. You don't want that! If you want one of these units, it's better to house only five or six chicks in it.

In other words, the relatively inexpensive brooder kits are usually the way to go with a small home flock. And even using a roomy (free) cardboard box (like an appliance box) for a DIY brooder is usually better than an expensive commercial brooder if you're just raising a small flock at home.

Here are the basic requirements for a brooder enclosure.

Egg-cetera!

fresh eggs

If your hen has sustained damage to her reproductive system (such as an infection in her ovaries), she may begin to look more like a rooster—and act like one—due to a hormone imbalance. At her molt, she'll grow rooster plumage. Once the condition clears up, she'll grow in henny feathers again.

A secure enclosure. How secure? It will depend on your situation. Baby chicks are delicate and need to be protected from other pets (like dogs and cats) and rodents—and they may also need to be protected from the unsupervised visits of well-meaning toddlers. If you have none of these in your house, all you will really need to do is make sure the chicks can't get out and get too cold or hurt themselves—minimum security. Where house pets, rodents, or little ones are a concern, you may need to add an extra layer of protection (or two) for safety. If your dogs or cats are especially unruly—or if you're worried your toddler will be able to access the

4 Things to Know about
Leaving Your Chickens Home Alone

Caring for chickens is so simple that you can leave them alone for a few days so long as you see to a few basic needs.

1. **They need enough food and water for the duration of your trip.** That should be a no-brainer. But more than that: If you're gone for more than a few days, you'll probably want to have someone just peek in and check on them to make sure they haven't, say, scratched their waterer full of shavings or overturned a feeder. If you leave them plenty of food and water but they spill it or can't get to it, it will do them no good.

2. **They need to be secure from predators.** Of course, this should also be a no-brainer, but sometimes people are tempted to think, "They'll be okay for two nights." Maybe. But maybe not. And how awful would it be to come home to find that your entire flock was slaughtered in a midnight raid by a raccoon? It only takes one night. Heck, it only takes 1 hour. So, if your run isn't secure, don't tempt fate by leaving the coop open and risking a predator attack while you are gone. If their outdoor space is not secure, you may need to leave them locked in their coop while you are gone, but, as you'll see, this is not always ideal. . . .

3. **They also need enough space.** If your chickens normally get ranging time but they are instead shut inside a small coop with nothing to do while you're away, they may develop some behavior problems. For instance, they may pick on each other or begin eating the eggs they lay. If you have to leave them in a small coop or small coop-and-run area, you may try to add a few distractions to the coop to keep them occupied. There are a host of chicken toys, treat balls, and snack racks on the market that fill the bill.

4. **Arrange for someone to gather eggs for you, if possible.** Neighbors, friends, and family members often enjoy helping gather eggs in your absence, particularly if you offer to let them keep what they gather. If you don't have anyone willing to gather for you, don't worry. If you're gone for just a few days, the eggs your hens will have laid during your absence should still be good to eat.

chicks—you'll have to make your brooder a maximum-security area.

Protection from drafts. Drafts will chill your birds, which can cause a variety of serious problems. This means that when you're looking for a brooder enclosure, you don't want something like a small animal cage—any drafts will blow right through it!

Plenty of space. Whatever housing solution you go with, make sure it provides 2 square feet per chick. It sounds like an awful lot, but as they get older (and bigger), you'll realize why this much space is necessary.

A heat source. Probably the worst mistake novice chicken keepers make with baby chicks is to keep them at regular room temperatures. Baby chicks will quickly die of hypothermia at those temperatures, so this can be a fatal mistake. Instead, when they've just hatched, chicks will need access to temperatures of 95°F in their brooder. (In fact, My Pet Chicken ships small orders of chicks with heating packs to make sure they stay warm along the way to your house.) A 250-watt infrared heat lamp, hung above one side of their living area to provide directional heat, is the most common way to keep chicks warm.

But the easiest way to provide the proper heat for your chicks? In our opinion, this would be the Brinsea EcoGlow brooder (or any similar product,

should one be developed). Let us explain why we like the EcoGlow best. There is nothing wrong with a heat lamp. They've been used for a long time. They're cheap and easy; at this time, you can get a heat lamp and bulb for $20 or less. If you need a stand to hang it on, that will cost a little more.

The problem with heat bulbs is that *if they're not hung securely and safely*, they can present significant fire hazards in the dusty environment of the coop and brooder. Especially in a small brooder—and if you're raising a backyard flock, your brooder will probably be relatively small—it can be difficult to get just the right temperature with a heat lamp because you have to adjust it by raising and lowering the lamp. If it's hung or aimed clumsily, a heat lamp can melt plastic feeders and waterers, or even the plastic storage container you're using for your enclosure. And, even worse, it can catch your cardboard box or bedding material on fire.

By contrast, the Brinsea EcoGlow is like a small plastic tunnel under which the chicks can huddle for warmth when needed. It doesn't get as hot as a heat lamp, and it doesn't present the same risk of fire that a 250-watt bulb does. It won't melt anything. The EcoGlow provides a small, warm area but also allows your chicks to escape the heat when needed. It doesn't heat up the whole room like a heat bulb does, meaning

Egg-cetera!

Mother hens can show their babies what is good to eat and what is not. In tests, hens have taught their babies to stay away from color-coded grains that are bad for them.

that if you have to keep your chicks in an area you use, the room won't be as hot for you (which could be good or bad). It also doesn't keep the brooder lit up 24 hours a day, allowing your chicks to get used to the dark and have natural periods of sleep and wakefulness. In short, it's a lot more like a mother hen than a heat lamp is. The only drawback is that these units are more expensive. On the other hand, if you don't have a good way to hang your heat lamp safely and you're planning to purchase a stand, the difference in price between a heat lamp and a small EcoGlow is probably marginal, at best.

If you do have to go with a traditional heat bulb, we recommend a red heat bulb for a few reasons. One, white light is brighter, so it can be especially hard for chicks to sleep with a white heat bulb. Two, they are less likely to peck at and hurt one another in the dimmer red light.

A chick waterer. Invest in a special chick waterer. Don't try to use a dish, a small dog bowl, or anything you just happen to find around the house. Why not? Baby chicks, bless them, can be clumsy. If they are perching on the side of a dish that is not designed for them, they can overturn it, get trapped beneath, and die. In other cases, the outcome may not be that dire, but they can still get wet and cold or just soak their bedding. Plus, they're babies; sometimes they'll just topple over, asleep. If you have a dish that is too large or deep, you're likely to have an accidental drowning. A rabbit waterer isn't preferable, either, because not enough chicks can access it at once (and they may not be able to even figure out how to use it). If you are in an emergency situation and have no other choice, use a shallow, wide-bottomed bowl that will be hard to overturn, and fill it with clean marbles or polished stones. The water level should just barely exceed the height of the stones. The chicks will be able to drink from the interstices, and there will be less danger of drowning or overturning. For the best results, though, we recommend you use a chick waterer. They come in a number of different sizes and shapes, all basically sufficient. Even with the best waterer, your chicks will still kick bedding into it and find ways to poop in it. Raising the waterer off the ground somewhat will help (starting in their 2nd week of life), but, no matter what, they're going to get that water messy, so plan on changing it a few times a day.

A feeder. Once again, we recommend you resist the temptation to use a dish or bowl for feeding your chicks. Your little charges are messy, and they'll jump in and kick the feed all over the place and poop in it. Plus, just like with a nonstandard waterer, a chick could get trapped underneath. Good feeders are not expensive at all, so just spend a few extra dollars and buy a genuine baby chick feeder. We find that hopper-style feeders, where the feed is refilled at the top and comes out the bottom, are usually a little easier to use than other models. As the chicks eat, gravity automatically refills their trough from the hopper. Trough-style feeders, while very inexpensive, are a big pain. They have to be refilled more frequently because there's no hopper. And the feed tray is usually so deep that chicks can't reach the feed at the bottom, when levels are getting low. Furthermore, chicks love to roost on them and, of course, poop in them. Hopper-style is the way to go.

Siting the Brooder

Deciding where to site your brooder or broody coop is a process similar to siting the "big girl" coop: There is no one perfect place to put it; it will depend on your personal situation. Broody hens can raise chicks outside in the coop, separate from any other flock members (or else in their special "broody coop"). It's sometimes possible to have your broody raise her chicks in the house, too, if you have a large enough brooder to fit her. Broody bantam hens are a popular choice to keep inside if you have small numbers of chicks to raise.

When you're using an artificial brooder, it's usually located inside the house because it's easier to provide warm, draft-free, secure conditions. An extra bathroom is a popular location if you are lucky enough to have one. That makes cleanup a breeze, because cleaning and sanitizing feeders and waterers is easier with a sink nearby. A laundry room with a utility sink is popular, too. However, there are plenty of people who use an extra bedroom or home office—or even a heated garage or basement for a brooder site.

Keep in mind that if you have to locate the brooder in an area of your home that is used frequently, like a bedroom or office, you *will* have dust. And we mean DUST, with a capital *D-U-S-T*. Of course, this will be true even if you have the luxury of an extra bathroom you can dedicate as a brooder, but if you have that dust-making factory in a home office, you may find it more of an imposition. In such an instance, you might consider using some old sheets to drape over furniture to keep the dust off. If you don't have any old sheets you can use for that purpose, getting some inexpensive drop cloths or plastic sheeting from your local hardware store is a good idea.

Or, of course, you can just keep dusting—but it does get tedious.

The world won't come to an end, of course, if you have to place your brooder in a bedroom or another room without the convenience of its own sink. But the world *will* (more or less) come to an end if the brooder is in an area not secure from your cats or rodents, or if it is in an area that gets drafts or overheats in the sun. Safety and security come first; convenience after!

Please consider these questions when deciding where to site your brooder.

• Can you locate the brooder in an area secure from unsupervised small children? Can you locate it in an area that is secure from pets?

• Can you locate it out of direct sunlight?

• Will it be easy to cover securely?

• Can you locate it out of drafts or breezes that might come from an open window, a fan, an air conditioner, or a heater vent?

• Can you insulate it from very cold surfaces (like a basement floor)?

• Can you locate it in an area where it will be convenient to hang a heat lamp?

• Will it afford a convenient place to plug in your heat lamp or EcoGlow?

• Can you locate it in an area you can easily and conveniently monitor?

• Can you locate it in an area that makes cleaning simple (with easy access to a sink, for example)?

• Can you safely secure a heat source so that it won't present a fire hazard?

Let Your Hen Raise Those Chicks

Having our chicks raised by a good broody hen is our favorite way to care for them. It's simply wonderful to see a mother hen with her babies—it's heartwarming when they ride on her back and peek out from beneath her wings, and it's adorable the way she teaches them about what is good to eat and how to scratch and forage for food.

It's practical, too. When a hen is brooding chicks, they won't need a heat lamp or EcoGlow heater because the hen will keep them warm. You needn't worry about how to provide heat if the power goes out or about accidentally keeping your brooder too hot or too cold. There are no concerns about the fire hazard of having a heat lamp in the brooder. And, finally, it lets your hens express an important natural instinct. Having chicks raised by a broody hen helps increase the chances that they'll be healthy and happy, and it also seems to help them integrate into the flock much more easily when it's time. The mama hen will help teach them about the pecking order and appropriate social behavior in your particular flock.

Even though having a broody hen raise your chicks is a great way to begin, it won't work for everyone. If you're just getting started in the hobby, you won't have a broody hen on hand to raise your chicks. If you do have or can acquire a broody hen, there are still some things to consider.

Broody Considerations

First, in order to raise chicks, your hen needs to be broody. Right now. And by that, we don't simply mean that she needs to be from a broody breed. Instead, she needs to be *currently experiencing the hormonal condition of broodiness*. In addition, not all broody hens are good mothers. If you want a broody to raise your chicks, you need a broody who *is* a good mother. Otherwise, she won't adopt them, and she may even try to kill the new chicks. When a hen is not brooding, her instinct is to protect her territory from interlopers, and she will see the chicks as creatures who are invading her territory, not as fluffy little peeps who need someone to care for them.

That also means that when she's hatching, you'll want to give your broody a special, safe place to incubate her eggs—a place away from the rest of the flock since your nonbroody hens may not take to her babies so calmly. Even another broody might see someone else's chicks as competition for her own potential brood. Although mama will try to protect the chicks from aggression from other flock members, she won't always be successful. In the old days, losing a chick or two in this way every so often on the farm was just the way things worked. It was just nature, right?

Well, it's still nature. But your small pet flock is a different situation from a traditional farm. On the farm, there was room for hens to keep their babies away from more aggressive companions. And there was space for babies to escape when they were being chased. Plus, if any chicks were lost, for the most part, it was relatively easy to hatch replacements right on-site. By contrast, in a backyard chicken scenario, you're probably not going to have the type of room hens had back in the day, so your mother hen and babies will not have as much space to avoid rabble-rousers. If you lose chicks, you probably don't have a rooster, either, so your own eggs won't be hatchable. You'd have to seek out more eggs and/or more chicks—only to face more possible losses.

So even when you're using a broody hen to care for baby chicks, you'll still need a separate safe place to keep them all, just as (without the hen) you would otherwise need a brooder. Using a broody hen isn't a real way to save on special equipment. It has many advantages, but if you're looking to save money, brooding by hen isn't really the way to do it. All you'll really save is the amount of money you would otherwise have spent on a heat lamp.

If you have decided to use a broody hen, however, it's important to choose the best for your hatching needs. Some breeds will never go broody, so if you are waiting for your favorite hen to get in the mood for hatching, it may be a very long wait! Our favorite broody breed is the Silkie (a bantam), but other popular favorites include Cochins, Orpingtons, and Old English Games.

Broodies and Shipped Chicks

An important note if your broody hen will be raising shipped chicks: We recommend allowing your new chicks 6 or so hours in a heated (about 95°F) brooding area first, with access to chick feed and water, before introducing them to your hen to be cared for.

Here's why: When hens hatch eggs on their own, they hover and continue to set for a while before taking the babies to eat and drink. In fact, their natural instinct is to wait for about 48 hours before they cease egg-sitting mode and begin chick-rearing mode, moving from the nest. The reason it takes a while for hens to break out of the egg sitting trance is that the chicks need time to dry off beneath her, and all her eggs need sufficient time to complete hatching. She can't jump up after the first egg has hatched. This would doom the rest of her clutch. She must wait until they are all done, so her instinct is to stay on the nest about 48 hours. The chicks absorb the egg yolk into their abdomens right before they hatch, so they don't have to eat and drink right away while they are waiting. It's a great system.

However, with babies who have been shipped to you, it is important to give them a little time to eat and drink first, before introducing them to momma hen. Shipped chicks won't arrive at your home for a day or two after they've hatched, so your chicks will have already used some, if not most, of their reserve. Mother hen won't know that, though. She treats the chicks as her own, as if she had just hatched them, which means it may be some hours before she encourages them to eat.

So when using a broody to raise your shipped chicks, give your babies a little time, 6 hours or so, to recover in their brooder. This way they will be sated and have time to wait until mother hen's instincts to leave the nest kick in, as many as 48 hours after she first feels the babies beneath her.

PREDATORS and PESTS

One of the big myths of chicken keeping is the old chestnut
that chickens will attract wild animals to your neighborhood.
It's not true; it's not even common sense.

Presuming you keep everything clean and tidy with your pet chickens (just as you would when keeping a pet cat or dog), raising them doesn't make wild animals magically appear from thin air. While it *is* true that pests that are already in your neighborhood might be attracted to chicken feed if you spill it or don't keep it secured, they would be just as attracted to spilled or unsecured cat or dog food, wild bird feed, a koi pond, or even your family's food waste discarded in unsecured outdoor garbage cans or compost piles—and all of these things probably already exist in most neighborhoods, anyway.

Chickens are also generally more vulnerable to attack by smaller predators than dogs or cats are—nearly everyone seems to like a chicken dinner! But even so, small dogs and cats that live in areas with predators are attacked by hungry wild animals, too. This doesn't mean small dogs and cats "attract" predators to the neighborhood, any more than chickens do.

Further, if you do have small rodents in your area, a flock of chickens can actually reduce their numbers, since some breeds will catch and eat small mice and moles like cats do; chickens will eat small snakes, too. Plus, they eat other pests like ticks, mosquitoes, grasshoppers, and other insects.

That said, when you are dealing with flock casualties, it can be helpful to identify what sort of predator you're dealing with, so you can take the appropriate steps to keep your flock safe.

So many My Pet Chicken fans and customers have e-mailed us photos of their dogs and chickens snuggling up. This photo was submitted by Rachel from Maine and features her Splash Silkie chick perching atop her dog, Lance.

DOMESTIC DOG

Domestic dogs—including those belonging to you as well as other people—are the most common predators of chickens in both suburban and rural areas. Dogs *can* be socialized with chickens (tens of thousands of chicken keepers have done it successfully across the country), and many even act as guard dogs, protecting their flock from predation.

Most dogs that attack chickens are not aiming to kill. They simply want to chase, but even chasing can be fatal because chickens often break their necks trying to get away or die of heart attacks if they have nowhere safe to escape. In cases where your chickens are neither hurt nor killed by a dog attack, they can still be thrown off laying for days or even several weeks due to the upset.

If you're unsure which predator has hurt your flock, the scene of the crime might lend a clue. If dogs are the culprit, you'll often find one or more chickens—sometimes an entire flock—laying scattered about with broken necks and just an occasional bite mark. Some dogs will simply chase them until they die; some will shake a chicken to death; some will kill with one hard bite and move on to the next. But well-fed pet dogs will rarely actually eat a chicken. They chase and kill just for sport. Many wonderful, sweet, calm, and gentle dogs—dogs who would never think of hurting a human being—have chasing and

Egg-cetera!

Chickens have a ZW (rather than an XY) system of sex determination. The sex of chicks is decided by the female within her ova, not by the male's sperm.

killing instincts when it comes to chickens.

Dogs will jump and even climb fences to get to chickens; they will dig under barriers and tear through chicken wire. If your dog kills your chickens, remember that it is not your dog's fault; he is responding to natural instincts. It is your responsibility to socialize your dog with your flock.

If someone else's dogs have hurt or killed your chickens, remember that in most areas, owners must keep dogs leashed or fenced, and are legally and financially responsible for any damage done by dogs that escape or are not properly confined. Legal consequences vary—check your local laws. In most cases, though, the owners will have to reimburse you for any damage the dogs caused to your coop and run, as well as for the loss of the pet chickens. (Most chicks cost just a few dollars each, but some exceedingly rare imported varieties sell for as much as $100 per chick. When it comes to estimating the value of adult birds, we wouldn't expect the rarest breeds and varieties to sell for less than $100 each, shipping included—if you can even find a breeder that will sell them. Even common breeds kept just for egg laying might be as much as $50 each to replace, including shipping.)

Please note that stray or feral dogs that are hungry and killing to eat will kill more like a coyote, so be sure to read that entry later in this chapter for further information.

HOUSE CAT

Domestic house cats very rarely attack adult chickens, but they are usually a serious danger to baby chicks, and occasionally to juvenile birds or small bantams. Some cats actually flee from chickens, even from bantams! That said, occasionally we do hear from people whose cats seem interested in their adult birds. Since many of the most common chickens are just as big as house cats, this is pretty rare. Cats are carnivores, sure, but they prefer to prey on very small birds, rodents, small rabbits, and other little creatures—not your 8-pound hen.

If a cat gets into the brooder, it may kill or hurt several chicks at once just from having trod on them in an enclosed location—and if that cat is yours, she may actually bring you one or two as a "gift." Not fun! However, feral cats that get access to your chicks in the yard may eat what they have killed, often starting with the head.

Having cats around adult birds is a good thing in most cases: Cats will hunt and kill any mice that are attracted to fallen feed, thereby helping to keep away other predators like snakes or weasels that may be attracted to the same prey. Just be sure to keep your baby chicks and smallest birds secure. A cat will stick her paw into a brooder and try to snag and pull out your babies, so use a secure screen or fine-mesh hardware cloth. Make sure your cat can't overturn your brooder, either, or lift the lid and nose in! Cats may want to jump on top to investigate, so make sure your brooder and heat lamp are very securely seated. We suggest you keep your chicks in their brooder away from your cats entirely, perhaps in another room.

RACCOON

For most people in the United States, raccoons are probably the worst and most common wild predator of chickens. You may think raccoons are small and cute—but if they can get to your chickens, your thinking will probably change. For one thing, raccoons aren't especially small. They can weigh up to 20 pounds or so. For another, the carnage they leave behind will permanently destroy the idea that they are cute.

A raccoon that gets into your coop or run will normally kill multiple birds if it can get to them, and the bodies will usually be left where they were killed, rather than carried away. Raccoons won't eat the whole bird or even most of it: They will often just eat the contents of your birds' crops and occasionally some of the chest. They are awfully destructive.

They can climb walls and over fencing, and they can reach their hands through wire mesh that is too small for their whole bodies to fit through. (In the last case, they will pull out whatever parts of your birds they can reach.) They can chew through aviary netting. They can dig under fences and into runs. They are smart—many latches do not adequately protect against raccoons—and, hunting in teams, they display a remarkable ability to plan and coordinate their attacks.

Raccoons are chiefly nocturnal, so if you range your flock during the day, be sure to lock your birds up securely at night with the precautions recommended later in this chapter in the "Securing the Coop" and "Securing the Run" sections. Raccoons are carriers of rabies, so if you do see a raccoon out during the day, be very cautious. It might just be hungry, but it might also be sick. Your birds can't get ill with rabies, as the avian metabolism is just too different—but you can! Use extreme caution.

OPOSSUM

Opossums are omnivorous marsupials that resemble a large rat that's more the size of a cat. They are gray to silver with many sharp teeth, and their long, hairless tails are semi-prehensile. They're on the small side, weighing up to about 14 pounds—but on average they tend to be half that size.

An opossum that gets into your coop or run will target eggs and young chicks, but they are certainly known to kill adult chickens as well. Small bantams are especially at risk. When opossums kill adult chickens, they will take one or more and leave remains quite similar to what a raccoon leaves—the body or bodies will usually be left where they were killed, rather than carried away. The birds will usually be killed at night by bites to the neck, and the

(continued on page 86)

Egg-cetera!

Some people worry that gathering eggs will upset their chickens. However, unless a chicken is broody, she will not be interested in or protective of the eggs she lays.

Training Your Dog
to Accept Your Chickens

Dogs can be a serious danger to your pet chickens. It's natural for them to chase small animals like chickens, but even "just chasing" could seriously injure or even kill your birds. So what should you do if you have a dog but want pet chickens, too?

Don't panic! You can have both. Remember, just because your dog's natural inclination is to chase or hunt doesn't mean you can't keep chickens. After all, housetraining is a learned, not instinctual, behavior for dogs. It can be done, and many people have both dogs and chickens. Just like with any sort of training, teaching a dog to get along with chickens will take time, effort, and consistency. Don't expect to see overnight success.

Here's how to do it: First, remember that during training sessions, you should be in a calm, easy mood, with an all-the-time-in-the-world attitude. Be completely focused on the task at hand. Look at your dog when you give him commands (even if he is not looking at you like he should, as the trainer, you must make eye contact when you give commands). Also, make sure that training sessions don't occur at the time of day when your dog is most playful and active. Your dog should be calm when you start to teach.

Your dog needs to know two important commands. The first is "leave it." This command is taught by laying an item on the ground—it can be anything, like a soda bottle, but it shouldn't be too tempting. When the item is on the ground, look at your dog and command (using his name), "Rover, leave it." If he tries to take it anyway, push him back, firmly but gently, and looking at him, say again, "Rover, leave it," in a stern tone. He may try repeatedly. Each time, push him back and repeat, "Rover, leave it." Eventually he'll get it.

This process may have to be repeated over many sessions until you no longer have to physically push your dog away from the item, and he accepts your authority at once. As he gets better at accepting your direction, you can proceed to items he will be more interested in, like a toy or a treat. These require more self-control on your dog's part.

The second important command is "be easy." "Be easy" is often preferable to a simple "no," because "be easy" can be used anytime you need your dog to settle down, whether he's excited by people, a knock, chickens, or something else.

Stage 1. The first training exercise is to take your dog on a 3-foot lead and calmly enter your chicken run. He may get excited or try to chase your birds, but keep him right by your side until he sees that he will not be permitted to continue his excited behavior. Pay attention to your dog's reaction and his attitude toward the chickens. If he doesn't settle down, or if you're worried about your ability to control him effectively, you will need to work with him *outside* the poultry yard for the first few sessions.

Using his name, remind your dog to be calm by saying, "Be easy, Rover, easy." Your tone should be smooth and calm, not sweet or coddling—but not aggressive or angry either. Be sure you have complete command and control of your dog at all times in case he makes a sudden lunge. The goal for the first few sessions is just to have him be there, without behaving in an excited way. If you like, grab a lawn chair and sit down; do not allow your dog to stray more than 3 feet from you. Your objective is to have him just sit or lay down calmly next to you. Once you have him sitting next to you quietly and peacefully for about 5 minutes, you have had a successful training session. You will want to repeat this lesson several times over the course of a couple of weeks, until you don't have to struggle to keep him calm and well behaved.

Stage 2. The next stage is to walk him on a lead, calm and easy, around the run. Start with sitting inside the poultry yard in a composed manner. If he starts to show excited interest in the chickens, correct him immediately. Your goal at this stage is to get your dog to walk calmly around the chickens while on a lead. There should be no play, lunging, barking, or chasing; he must learn to walk calmly around the chickens. Use the commands "leave it" and "be easy" as often as you need until eventually your dog assimilates the training.

Stages 3 and 4—And Beyond. The idea is to progressively give your dog more autonomy when he demonstrates good behavior and self-control. Once your dog is comfortable moving among your chickens calmly—without showing any inclination to jump, lunge, or chase—you can give him 10 feet, while you maintain control at the end of the lead.

When that is mastered, you can drop the lead

and let him drag it around in the run. Next, remove the lead entirely, but stay in the run with him. If your dog is calm and easy after an hour, then you can leave the poultry yard, as long as you are watching your dog for signs of excitement toward the chickens. When you are satisfied that he shows no inclination to harass your birds, you can stop watching for short periods of time, growing longer and longer until you are satisfied that your dog presents no danger to your chickens.

Important Notes. Training your dog requires practice and repetition; don't get impatient and rush things along. The lives of your chickens depend on getting your dog trained correctly, not on getting him trained quickly. At any stage, if your dog reverts to bad behavior around your flock, you have likely moved on too soon. Simply go back one step.

Finally, some dogs with strong prey drives will be especially resistant to training. Livestock guardian breeds, which are used to guard livestock professionally, have low prey drives for a reason. So keep in mind that in some cases, no matter what you do, your dog may be temperamentally unsuited to accepting chickens. If that's the case, the best idea is to keep your dog and your chickens safely separated.

Contributed by Jackie Church, owner of Windance Farms in upstate New York, where she breeds and raises Maremma Sheepdog livestock guardians. She has written an LGD Training Manual and also mentors, consults, and is available for seminars on how to raise and train successful LGDs. Visit www.windancefarms.com.

opossums often eat just the contents of your birds' crops and occasionally some of the chest. Sometimes they will just drink the blood—ugh! But in any case, they don't normally consume the whole bird.

Opossums are not as smart as raccoons, and they're not as adept at opening locks or latches. However, they can fit through very small openings for their body size—as small as 2 inches! Like raccoons, they can reach their hands through wire mesh that is too small for their whole bodies to fit through, and they will pull out whatever parts of your birds they can reach. Opossums can climb fences and dig beneath them, too, although they do move quite slowly and can even seem clumsy. This (along with the fact that they're nocturnal) is why most casualties will occur at night, when your birds are sleeping.

Birds of prey are, of course, a serious threat to chickens, but they can also help. Here, a sharp-shinned hawk makes off with a snake.

SNAKE

Most snakes in the United States are too small to prey on adult chickens. Instead, they tend to attack eggs and young chicks. Very small snakes won't even be a danger to your eggs, as they primarily eat insects, worms, tadpoles, minnows, salamanders, slugs, or tiny frogs. The smallest snakes are probably more in danger *from* your chickens than a danger *to* your chickens. Your chickens will likely hunt and eat small racers and garters, and may even regard them as a special treat. You may have the pleasure (or surprise) of watching one of your chickens catch a tiny snake and then flee the flock with the little snake dangling from her beak, occasionally having the special prize stolen by another hen, and generally playing "keep away" until someone has enough time to gobble it down.

Snakes that are large enough to be a danger to your eggs may first be attracted to your area because they are eating mice or rodents, so be sure to take care of any rodent problems to avoid attracting snakes to your yard in the first place. Further, be sure to find any hidden eggs your hens may lay in your yard and encourage your flock to lay inside their coop, where you will be sure to find their eggs. Gather your eggs regularly so meals won't be available.

If snakes are large enough to be stealing your eggs, however, you may see no signs other than the missing eggs. Even larger snakes can squeeze through very small openings, so they are difficult to exclude from the coop. After having eaten an egg, the snake may temporarily have too much girth to escape from the same opening it entered, if it was a tight fit. If snakes have grown accustomed to stealing eggs from you, they may even eat any wooden eggs or golf balls you place in the nest to show your chickens where to lay, and that may lead them to an early end.

Adult birds will probably not be in danger from most snakes, although attacks are not unheard of. If there are any remains in the coop, you might find a chicken with a wet head, indicating that the snake tried to swallow her and then gave up because she was too large—they generally want to start with the head. Since they ingest their prey whole, snakes that do

TRACI'S SQUAWKINGS

A couple of months ago, I dreamed my flock was destroyed. I bolted upright out of bed, and then, in a fog, realized it was just a dream. The coop in my dream was the wrong one—it was from a house my husband, Derek, and I lived in years ago. And we'd recently predator-proofed our run. I reassured myself everything was fine, and after some time, fell back asleep.

A few days later, Derek and I were up late, talking, with the windows open. Around 11:00 p.m. we heard a squawk. (Any chicken owner will tell you that a squawk at 11:00 p.m. is a *bad* thing.) We ran outside to find not one but *three* raccoons. One had made its way inside our reinforced run and nabbed—of course—our daughters' favorite chicken.

We scared the raccoons away, buried our friend, found the spot where they'd gotten into the run, reinforced it with more zip ties, closed the door from the coop to the run, just in case—and our nightmare was over. So we thought.

Around 4:00 p.m. the next day, while pounding away on our respective keyboards, we saw one chicken cross the road. Literally. As everyone knows, chickens *don't* do that. We should have recognized that something was wrong, but we'd never lost a chicken to a daytime attack, so instead figured we'd have some chicken-herding to do that night.

An hour later, Derek went out and found a bloodbath—mangled corpses, piles of feathers all over the place. My nightmare came true: Raccoons killed my chickens. Final tally: 11 chickens out of 31 killed.

Our best guess: Those raccoons were ticked that we stole their kill from them, woke up early, and tag-teamed to kill as many chickens as possible.

The thing that makes me angriest about raccoons is that they kill as many chickens as they can— many more than they could possibly eat—before taking off. Birds of prey take just one. Most foxes will take one chicken and run for it. But raccoons kill wantonly.

manage to swallow an entire chicken would not be able to get out of the coop again unless there is already a chicken-size opening in the coop. If the snakes can escape your coop after having dined on chicken, you will probably see no sign, so it can be difficult to determine the cause of the loss.

RAT

Rats will generally only prey on chicks and eggs, but if you have a serious infestation and food is scarce, they can also raid your coop at night and attack your juvenile and adult birds.

Rats kill by biting the head or neck, like members of the weasel family. Like opossums, they will drink the blood of their victims. Parts of your chickens' bodies may be eaten, and the corpses may be pulled into burrows or other concealed locations for feeding. Sometimes they are just dragged into a corner.

Rats carry disease, so it is important to control them if you spot any signs. Even if they aren't preying directly on your chickens, when rodents defecate in your chickens' feed when eating it at night, it's bad news. Your chickens may ingest the feces, and that can make your girls ill with salmonella or other diseases. You may not see any rats unless you have a severe problem. They are shy and come out mostly at night. However, if they are around, you may be able to see droppings or spot chewing damage on your coop. If your coop has no ingress for them, they will make their own, so keep an eye out for gnawed corners. They are also great burrowers and can enter your coop or run from below.

If you see gnawing damage on your coop, try stapling hardware cloth around the damaged area so the rats will not get through. Since rats are burrowers, you may wish to wire the bottom of your run, bury the bottom of your fence, or provide an apron so they can't enter your chickens' area from below. Since they can climb fences, though, the only way to exclude them from the chicken yard absolutely would be a completely enclosed run: top, bottom, and sides.

FOX

Foxes will not appear in suburban settings as often as raccoons or opossums, but they are still a danger, often because they are so cunning. They will normally stake out your coop beforehand so they know when to strike and grab a bird, and how to get away fast. Usually they will take just one at a time. That said, if they get inside an enclosed coop or run, it's possible that many or all of your chickens will be killed, usually with broken necks, and as many as possible will be carried away. Occasionally there

Egg-cetera!

Chicken bones found at El Arenal-1, an archaeological site in Chile, suggest that Polynesian sailors may have come to the Americas before Christopher Columbus's 1492 voyage.

will be only parts of birds missing if the fox can't get the whole bird out of the coop. Foxes may also get into your feed if it is not protected adequately.

When foxes are the predators of your chickens, you will find bunches of feathers pulled out and strewn about. Often you will find feathers outside the coop and run and along the fox's escape route, too. Survivors may have sustained injuries as they fled, including missing clumps of feathers near the neck (a close call!), wounded necks, and occasionally wounds near the vent.

WEASEL FAMILY (INCLUDING SKUNKS)

This group contains many different predators, including the weasel, ermine, ferret, mink, badger, and fisher; which ones you will see in your area depends on where you live. Their scientific names can be confusing, but all belong to the same scientific family, Mustelidae (commonly called the weasel family), except for skunks, which are in the Mephitidae family. (For convenience, "weasel family" will refer to skunks, too, in this section.) Those species interested in preying on your chickens generally are small- to medium-size, long-bodied, short-legged omnivorous mammals. Some are tiny; for instance, least weasels aren't much bigger than a mouse, but they can kill rabbits and chickens many times their size. Other members of this family are very large; for instance, wolverines may weigh up to 70 pounds.

Some animals in the weasel family (otters and groundhogs, for example) are not predators of chickens. Some, including the skunk, are more often predators of young chicks or eggs.

When a chicken has been taken by a predator, often the only trace will be a pile of feathers on the ground.

However, most animals in this family will eat chickens if given the opportunity, and when they do, their hunting behavior is similar enough to be classified together here.

Weasel family predators are often effective hunters. The lithe body shape of many of them means they can squeeze through holes that mice or other small rodents can barely fit through. In fact, most animals in this group usually prefer to prey on voles, mice, and shrews, so their body shape is quite efficient. Chickens are not the preferred prey of many members of this group. Larger members like fishers, wolverines, and badgers are an exception: They can be vicious, determined hunters of your chickens if they are in your area.

Weasel family predators are opportunistic. Since the primary diet of many small weasels consists of small rodents, a weasel who has gotten into your coop may initially have done so because it was hunting mice, or even because it found your unsecured chicken feed, hidden or ungathered eggs, or other attractive edibles. What tends to happen is that once they eat most of the rodents and preferred prey in your area, they may also begin targeting eggs, chicks, and adult birds.

Weasel family predators are typically excellent climbers and diggers, and larger types can be quite strong, forcing their way through the edges of wire mesh that aren't secured. If these are the predators of your chickens, they will probably kill multiple birds, with lots of small bites to the base of the skull. The larger predators in this group may carry off a carcass, but since they tend to gain access through smaller openings, it may not be possible to remove a carcass from the scene of the crime. Instead, the bodies may be strewn about where they were killed, or they may be piled. Which particular species did the damage can probably best be determined by learning which ones are common in your area. Skunk predators will, of course, leave a lingering smell.

BIRDS OF PREY

Birds of prey (the daytime hunting raptor from the Accipitridae family and the nighttime hunting owls from the Strigidae and Tytonidae families) come in many different sizes and colors, but all are carnivores and generally have sharp talons and strong, curved beaks designed for eating flesh. Unlike most land-based predators of chickens (which may kill many birds or an entire flock at once), flying predators will normally kill only one chicken at a time. Due to the predatory birds' long, sharp talons, the chicken killed may appear to have been stabbed with a knife, with many deep wounds and slashes, usually on the back and breast. The breast is often eaten. Owls will eat the head as well, while day-hunting birds will cleanly pluck feathers. Sometimes, however, if the bird of prey is large and your chickens are small bantams or young birds, they may just be carried off to be eaten elsewhere, with nothing remaining behind.

Birds of prey may defecate near the kill site; you may see stripes or splashes of white on the grass or soil or on buildings or trees. If you have a day-hunting bird, for example, you might find feathers plucked and scattered, and then stripes of whitewash from defecation on or close to a nearby fence post, tree, or telephone pole. Owl defecations will be chunkier and chalkier, and owls generally don't pluck feathers.

Many day-hunting raptors are migratory and will usually not stay in the area harassing your flock for long, especially if you live in an urban area. We tend to hear more anecdotal reports of predation by hawks in the spring and fall, perhaps when many of these birds are on the

Egg-cetera! *fresh eggs*

Chickens are believed to be the closest living relative of the Tyrannosaurus rex.

Birds of prey are among the most difficult predators to ward off. Here, a hawk that has been spooking our chickens opted for a toad instead.

move. If you live in a rural area, you may be located in the summer or winter hunting grounds of these birds. Even then, chickens are not their preferred prey.

The raptor that most often preys on chickens is probably the red-tailed hawk, because it is a relatively common raptor, and because it is large enough to prey on chickens easily. Bald and golden eagles and other larger birds may prey on your chickens, too. Even very small hawks like the Sharp-Shinned may sometimes try to get one of your chickens, although they normally stick to hunting small wild birds the size of sparrows and robins.

Do not hunt or trap raptors. They are protected by international treaty. If you hunt or trap them, you could be looking at genuine jail time or serious fines—sometimes even just for possessing a single feather (and even when you simply found that feather on the ground). Seriously, just do *not*. We can't stress this

enough. If you do have a raptor hanging around your coop for a long period of time and your birds are not in a secure, covered run, you can try calling your local county agricultural extension office to see if they will relocate it for you, or if they can point you to someone licensed to do so.

CROW AND OTHER CORVIDS

Corvids are large perching birds, usually with dark coloration, although some (especially tropical species) can be very brightly feathered. Crows are among the most intelligent animals, and can even make and use tools. They can also recognize individual people by their faces.

If they go after your flock at all, it will probably be an opportunistic attack on chicks, young birds, or small bantams. They may also be sneaky and try to steal eggs. They are unlikely to attack adult chickens. If you see a large bird carrying off a chick from afar, you will probably be able to tell whether it is a bird of prey (such as a hawk or owl) or a crow or corvid, because a hawk will carry off prey with its feet, while a crow will carry off prey in its beak. Corvids are not raptors and will simply not attack the way birds of prey will. In fact, the word *raptor* comes from a Latin word that means "to seize," and it refers to raptors' proclivity for seizing prey with their strong feet and talons.

Crows are fairly unlikely predators, even for your young birds or eggs. They are opportunists, and if they see an easy meal, they may take it. They may spot and eat eggs laid in the yard rather than in the coop, and in cases where eggs aren't gathered frequently enough, they may even learn to sneak into your coop to steal food or eggs from nests.

That said, generally speaking, if there are crows around your house or coop, it is good news for your flock. Crows hate hawks, so they will often mob up into a large group to drive away any hawk who makes the mistake of hanging around in their area. Crows recognize that hawks will prey on their eggs and chicks, so whole groups of them will harass the real predators until they leave.

If you live in town and don't have roosters around, the loud, scolding warning of crows can alert your hens that a predator is in the area. Some roosters will even learn to heed the warnings of crows—so if a hawk is out of sight of your coop, when the crows start calling, your roosters may herd the girls to cover until the crows stop and the coast is clear.

WOLF AND COYOTE

Wolves generally hunt in packs. Coyotes may sometimes hunt singly or in pairs. However, you will probably rarely actually see either of them take your birds. They work mostly at night, and they also tend to avoid humans. Wolves are larger: They can be up to 5 feet long (excluding tails), and nearly 3 feet tall at the shoulder. They can weigh up to 175 pounds, but average around 80. Coyotes can be up to about 3 feet long (excluding the tail), and about 2 feet tall at the shoulder. They can weigh up to 75 pounds, but average around 50. Even so, the evidence they leave behind, if any, will be remarkably similar.

Wolves and coyotes will eat what they kill, in contrast to domestic dogs that kill accidentally, or for the fun of the chase. These wild canids prefer to break the neck with their strong jaws, but they'll grab whatever they can of a fleeing chicken. They prefer to single out the old, the young, and the weak, so those will be taken preferentially. They have a good ability to distinguish the slowest or least aware in your flock, often birds whose vision is obscured by large crests.

When your chickens are killed by coyotes or wolves (or feral dogs), if you find remains, they

Egg-cetera!

Current research suggests that chickens have been domesticated for about 8,000 years.

Coyotes are wily and can easily get over fences as high as 6 feet. Some people have reported success repelling coyotes by applying wolf urine to the perimeter of their chicken yard!

will usually be disemboweled and mostly eaten. There may be puncture wounds from teeth in what remains of the body. However, chickens are small enough that they are usually carried off to be eaten elsewhere, so it may be hard to tell if your birds were killed by wolves, coyotes, or foxes. If your birds are stolen from a secure run, foxes will most often sneak in or climb over, rarely digging. If they do dig, foxes will dig smaller holes since they don't need as much space to squeeze through. Wolves or coyotes require a large opening to get in, so that might help you determine which of these predators you're facing.

LARGE FELINES

Canada lynx, cougars, and bobcats are large felines. Lynx can be up to 66 pounds and more than 4 feet long. They're generally golden or pale in color, with a very short, black-tipped tail. Their fur may be plain or spotted and lighter on the chest, stomach, and insides of the legs. They have long tufts on the tips of their ears, and a

ruff of long fur around their necks gives them fluffy-looking cheeks. Bobcats, relatives of the Canada lynx, are distinguished by their smaller size (up to 35 pounds and 3 feet long), darker brown or brownish-red spotted coats, shorter ear tufts, and relatively longer tail (up to 6 inches). Cougars, by contrast, weigh up to 220 pounds and can be up to 9 feet long, including the tail, with a plain coat that is tawny colored with white on the muzzle.

Lynx and cougars are uncommon predators of chickens in urban and suburban areas. Cougars especially tend to avoid areas of human habitation, but there are always exceptions. Bobcats, while elusive, are making a comeback in rural and suburban areas, and reports of attacks on chickens are on the rise.

In most cases, you will not find remains if the predator is a large feline; your chickens will simply be carried off. They will normally not kill more than they will eat, but again, there have been some exceptions.

If there are no remains, it will be hard to differentiate between a kill by wolves or coyotes and one by wildcats, unless you actually see the predator or its tracks. However, when they do kill more than they can eat, large felines will often shallowly cover uneaten remains with snow, dirt, leaves, or other detritus in a safe place so they can come back and feed again later. If you find such a spot, you can differentiate among cougar, bobcat, or lynx predators. Cougars can reach and scrape from 2 feet away to cover the remains, while the smaller lynx and bobcats will scrape from 12 to 16 inches away at most.

SNAPPING TURTLE

Snapping turtles are large freshwater turtles. And by large, we mean *large*. Common snapping turtles (*Chelydra serpentine*) weigh up to 75 pounds, but generally average around 35 pounds for an adult. Alligator snapping turtles (*Macrochelys temminckii*) have weighed up to 249 pounds in captivity, but in the wild *only* get to be as large as 175 pounds or so (is that all?).

Snapping turtles are infrequent predators of chickens, although if you have snapping turtles in your area, they may pose problems. Females especially are known to travel relatively far from their homes in ponds, streams, swampy areas, and estuaries. Snapping turtles can eat baby chicks and small birds, and although adult chickens are too large for most snappers to consider as lunch, the largest may indeed want your favorite hens. The larger turtles remind us of dragons—they are fiercely antediluvian looking.

Even small snapping turtles are notoriously grumpy on land, and may seriously injure the feet and legs of adult birds that get too near. One snap, even from a small snapper, can remove your finger or a chicken's toes or leg—be careful! However, this is merely a defense mechanism, since they are so awkward on land. If encountered in or near water, these reptiles are shy and normally just swim away. If you have to remove one—even a small one—don't ever pick it up by hand. If it's not too heavy, you can pick it up with a shovel and carry it someplace else.

Although they can dig, snapping turtles won't likely dig to get into your coop. They usually only dig in sand to lay their eggs (or down into mud in shallow areas to cover themselves), not to find a way into your chicken run. To protect your birds from snapping turtles, a simple sturdy fence will usually do. Snapping turtles don't climb or jump, and won't

Egg-cetera!

fresh eggs

The case of the Rooster of Basel occurred in 1474 when a rooster was tried, found guilty, and solemnly burned at the stake as a witch for "the heinous and unnatural crime of laying an egg." It's likely the poor chicken was simply a hen with a hormone imbalance.

(typically) hunt for chickens. They don't really have a method of getting into a secured coop. They will only be opportunistic predators, meaning that if they happen to come across a chicken while they're on their way somewhere else, they may take a snap, because who doesn't like a chicken dinner?

PROTECTING YOUR FLOCK

After just having read about all the hungry predators craving a chicken dinner, you may be worried if you range your flock in unsecured areas. You may be wondering: If the only way to prevent predator attacks is to keep your flock in a secure run and coop at all times, does this mean your ranged birds are doomed?

No, it doesn't. But it does mean that if you allow your chickens to range in unsecured areas, chances are that you *will* lose some to predators, over time. It can be devastating, especially to those with small, well-loved flocks. But we feel that the trade-off is worth it: Once you've seen how excited your flock becomes when they know they're about to be let out, it's really difficult to deprive them of that freedom. However, we can't fault anyone for wanting to keep their flock absolutely secure, either.

Securing the Coop

Your coop will need to be secure, whether you're ranging your birds or keeping them confined. Here are some measures to take.

Be sure your flock is locked in safely at night, when most chicken predators will be hunting, and when your chickens will be sleeping and at their most vulnerable. Your responsibilities in this area will depend on the management style you've chosen (refer back to Chapter 5 about this).

Use quality latches: It's not enough for the coop to be closed; it should be secure. Raccoons, for instance, have great manual dexterity and are smart enough to open complex latches and closures in laboratory tests. This means that latches that require two or more steps to operate are best. For example, a simple gate latch can be secured from raccoons usually just by inserting a screw-lock carabiner into the bottom hole. On the other hand, some people simply padlock everything. This is not usually necessary and can make access more inconvenient for you, too. However, with a padlock, you will know the girls are safe.

Don't give predators places to hide. A good choice is to use chicken coops that are raised off the ground, so weasels, rodents, snakes, and other small predators don't have good places to hide during the day (some bad guys can chew in from below).

Screen any ventilation holes or windows with a fine-mesh ($\frac{1}{2}$- or $\frac{1}{4}$-inch) hardware cloth. You don't want any predators sneaking in unannounced!

Check regularly for areas where predators may be trying to force their way into the coop. Some predators will try to force their way through small openings or gaps in wire-screen windows or vents; others may try to gnaw through walls or floors. Secure any part of the coop that's loosely attached; you might reinforce gnawed areas with wire mesh. The predators may be after your chickens, but they may also simply be after chicken food or eggs. Regardless, it's best to catch any threats early.

Securing the Run

You may or may not need to secure your run, depending on the management style you've chosen (refer back to Chapter 5 about this). The general rule of thumb is that the smaller the run, the more secure it will need to be. It's worth the trouble to do it right.

If you have a run that is *not* secure and a predator gets into it—particularly if the run is very small—*your birds will have no means of escape*. If they were ranging, or even if they had a larger area, they could run, hide, or fly away. Not so in a coop or small run where they may be trapped with any hungry predator that gets in. If that's the case, they may pile on top of one another in a panic and trample each other. So even if the predator doesn't make a successful kill—or if it only takes one bird and leaves—your birds may still get very hurt.

Consider choosing breeds that are quick enough to escape predator attacks in a range situation, if you will be ranging your birds in unsecured areas. For example, you might choose smaller "flighty" chicken breeds like Hamburgs, Leghorns, Campines, Anconas, or various game bantams. Easter Eggers and Ameraucanas are swift and alert, too. Some breeds have plumage that helps them avoid notice, like Welsummers and Speckled Sussex. You'll want to avoid breeds with large crests like Polish, Sultans, and Silkies. If you plan to keep more vulnerable breeds, you'll most likely want to keep them secured—or be prepared for losses.

If you will be constructing a secure run for your flock, use hardware cloth. A properly constructed run using fine-mesh hardware cloth will keep out both large and small predators. Some common predators can reach through fence openings with paws or claws and pull parts of your chickens out. Some, like weasels or snakes, can simply fit through the mesh of your fence, as if your fence isn't even there. Other predators may be so large they can force their snout or part of their head through an opening. For all these predators, the solution is simple: Use a fine-mesh ($\frac{1}{2}$- or $\frac{1}{4}$-inch) hardware cloth to create your runs and to secure openings. This way, even smaller predators can't squeeze or reach through.

Egg-cetera! fresh eggs

It's thought that chickens have about 30 distinct vocalizations that they use to communicate with each other. For instance, the warning cry for a predator that is coming on the ground is different from the cry for a predator coming from above.

Even the craftiest of raccoons has never broken into a door secured with dual barrel bolts like those pictured here.

For digging predators, either bury your fencing around the perimeter (a foot deep or so) or create an apron. If you have a very small run, you might decide to simply cover the bottom of your run, too. If you want to keep rodents out of your run entirely, which we highly recommend, either bury your fencing at least 24 inches deep, or, if you can't dig that far down, bury it 12 inches deep and create an apron around the perimeter of a minimum of 12 inches wide.

For protection from above, think about opting for a covered run. No tall fence will keep out flying hunters or climbers, so the only way to exclude them is to cover your run. This is a complicated issue though, particularly if you live in a snowy climate: The stronger and more predator-proof your covering, the more likely leaves and snow are to collect on it and, ultimately, destroy it. So we recommend you either:

• Make a predator-proof fortress of your run using fine-mesh ($\frac{1}{2}$- or $\frac{1}{4}$-inch) hardware cloth and sturdy posts at the perimeter, and supports in the middle if necessary, so neither predators nor snow and ice can damage it.

• Or use a lightweight, inexpensive choice like deer or aviary netting to cover your run, loosely, just to keep birds of prey out, and couple this with an automatic chicken door that closes your flock in securely every night, without fail.

In the former case, if it is truly predator-proof, you can leave the door from your run to your coop open at all times, meaning your flock can have total freedom to come and go as they please. In the latter case, however, or if you choose not to cover your run at all, you'll need to diligently open and close the chicken door every morning and evening. Or get yourself an automatic chicken door. Automatic doors come in both solar-powered and conventionally powered models; some sense dawn and dusk while others require you to set the time at which they open and close. Either way, they're an investment well worth your while, and one we recommend universally to those who can afford it.

With a large run, consider providing partial rather than full cover. This gives your chickens a place to hide and take cover from predators attacking from above, but does not provide complete protection. (In other words, if your chickens don't spot the hawk, they can still be taken by surprise.) Even providing a row of bushes for your chickens to hide beneath near the coop offers tremendous help for them in the case of flying predators. Birds of prey can't effectively swoop down through branches, and your chickens will (hopefully) still be able to see through them to determine when it's safe to come out again.

Check regularly for areas where predators may be trying to dig under your run or force their way past insecurely attached wire. You may need to restaple or screw in your fencing. In some cases, you may need to replace an apron or rebury fence. For areas that have been excavated, you might also choose to backfill with crushed rock or gravel, which will be more difficult for predators to dig through than just regular soil. Whatever the problem may be, your best bet is to catch it early!

Other Precautions

Securing your coop and run is a good first step, but good caretakers are proactive and will take some additional precautionary measures to help ensure the safety of their flocks.

Will your setup be secure against predators and rodents?

Take steps to avoid attracting predators or pests to your yard in the first place. For instance, make sure there are no food sources in your yard or area. Don't feed your cats or dogs

Egg-cetera! *fresh eggs*

Different chicken breeds have different voices. Faverolles have a distinctive, pretty warble and tend to be chatty. Rhode Island Reds say very little except when they have just laid an egg. And when Silkies are nervous, at times, it can almost sound as if they're laughing.

outside, unless you make sure you've cleaned up any leavings. Don't throw food scraps into an open compost pile. Make sure wild bird feeders are secure from rodents and other pests; clean up spills of seed. Don't leave out sources of water. In dry areas or during dry periods, animals may be attracted to sources of water in your yard, too. Don't leave anything out for them!

Be sure to store your chicken feed in predator-proof containers. Rodents are more likely to chew through wood than plastic, and they are more likely to chew through plastic than metal. If they discover a food source like your stored feed, they will be difficult to get rid of. Store your feed in a clean metal trash can with a tight-fitting lid.

To discourage smaller pests, you may want to remove your flock's feeders and waterers each night and replace them in the morning. While your chickens won't want to eat or drink at night, rodents, raccoons, and other pests most certainly will! Especially if your feeders and waterers go outside your coop, you'll want to make sure they're not attracting unwanted diners.

If you live in a place where roosters are allowed, be aware that they are great in helping to keep your chickens aware of dangers from hawks. In fact, they have a specific cry to let the flock know to take cover from an airborne danger. While the hens are foraging for tasty goodies, the rooster will keep an eye out for danger, and he will place himself in harm's way to give the hens time to escape to safety.

Use poison with real care—if at all. We recommend staying away from it entirely. You don't want your chickens getting into the poison, and you don't want that risk for your other pets and family members either! You also don't want your cats or dogs getting hold of a poisoned mouse or rat and getting poisoned secondhand. And you don't want a rat or mouse dying beneath your house or coop or in a wall, and decomposing (stinkily) there. Avoid attracting predators to begin with, and if you do encounter a problem, traps and exclusion are usually the best tactics.

Starting and Caring for Your Flock

Acquiring any pet is a big responsibility, and chickens are no different. The vast majority of My Pet Chicken's customers come into the hobby having thoroughly researched and educated themselves. By the time they place a chick order, most of our customers already know the basics about how to care for their new pets. They're prepared with a brooder, coop, and run, and they understand what to expect when it comes to chicken behavior. During the online checkout process, our customers are required to affirm that they will provide their chicks with proper care before they can even complete their order—and we provide free care information on our web site, just to be sure our chicks will be going to homes that are prepared to receive them.

In this part of the book, you'll learn how to become chicken savvy, from planning for your baby chick order (there's more to it than you'd think) to managing your adult flock.

ESTABLISHING Your FLOCK

There are still some surprises with chickens that even people like you,
who have spent serious time and energy to properly prepare, are not primed to
handle. We want to make sure that you're at ease with the whole process.

You may have read everything you can get your hands on, but other how-to books don't adequately explain how the ordering process works, or what your responsibilities are when it comes to retrieving your chicks promptly at the post office. You may not know enough about the process to even consider asking about what happens if there is a hatch-day issue with your chicks, a sexing error, or any number of things that can affect the fulfillment of your chick order. (Hatcheries are literally in the business of counting their chicks before they hatch, after all, so it's important to be flexible!) In this chapter, we'll teach you how to plan for chicks like a pro and what to do if your chick order is not able to be shipped as planned.

16 ROOKIE MISTAKES TO AVOID WHEN ORDERING

1. **Don't place your order without first having educated yourself about care.** Kudos to you if you're reading this book to avoid that pitfall!

2. **Don't place your order before checking to make sure it's legal to raise chickens in your area.** You might be surprised at the number of people who don't bother to check if they can legally keep chickens. It also happens that you may check with the zoning board, who tells you chickens are okay, only later to find out the health board or your housing association forbids them.

Conversely, sometimes people assume it's not legal to keep chickens, when it is perfectly fine. Do your research!

3. **Don't place your order until you're sure you won't change your mind about breeds, quantities, and ship dates.** In some cases there are fees associated with changing orders, and they can add up if you change your order repeatedly. But that shouldn't be the only reason you want to place your order just once. The more changes you make in your order, the more likely it is that a mistake will be made somewhere along the way. It could result in breeds or a ship date you didn't want. Our hatchery has the most advanced order-reserving software of any in the nation. Many other companies are still entirely paper-based, and errors will be more likely.

In addition, the birds you have reserved can't be adopted by anyone else, so if you've reserved them for yourself and then decide you don't want them after all, they must go up for adoption again. Remember, the chicks have to ship on the day they've hatched, so if you decide you want six Faverolles for March and then later decide you'd rather have six in May, your hatchery will have to find homes for the March babies that you rejected.

4. **Don't place an order for a "hatchery choice" assortment unless you're sure you'll be happy with any of the breeds you might get.** Depending on the assortment you order, you may end up with breeds that aren't cold hardy or heat-tolerant enough for your area. Or you may want birds primarily for egg laying, but end up with fancy pet breeds that don't lay a lot of eggs. If you have specific needs, place your order for specific birds. Many hatcheries, like ours, will allow you to choose alternates or substitutions if you do have some flexibility.

5. **Don't place an order until you've read the cancellation and change policies of your hatchery.** And don't place the order unless it will be okay if those rules apply to you. At My Pet Chicken, before you can place your order, you are required to affirm that you've read and agreed to the cancellation and change policies. Other hatcheries may not have a system that makes it as easy to access and understand their policies, but that doesn't mean you shouldn't be proactive. Then, after you have read the policies, think to yourself, "If I have to change or cancel my order, would I

Chicks need TLC, so make sure you're prepared to care for them.

be okay with the way they handle changes and cancellations?" If you wouldn't be happy, check out the policies of some other hatcheries. You may find one you think is more fair, or else you may come to learn that the policy you didn't like is an industry standard, and is actually fair after all.

6. **Don't place an order until you understand how losses or errors are handled.** This is similar to cancellation and change policies. Be sure to *read* the policies beforehand, shop around if you are unhappy, and consider whether or not you'd be okay with the way things are handled if you experienced a loss or an error. Remember, you can't send a baby chick back if there's been a sexing error. If you lose a chick, in most cases your remedy is a refund, not a replacement, since it's not safe to ship just one chick at a time.

7. **Don't assume that an order for chicks includes anything but the chicks.** They don't come with a brooder, feed, feeders, waterers, or other necessary supplies—you must have those prepared before the chicks arrive. Likewise, if you're ordering your brooder supplies at the same time as your chicks, make sure the supplies will come first so you'll be ready for the chicks. Remember, the hatchery can't ship your chicks in the same box with a heavy bag of feed banging around. The feed and the chicks may not even originate from the same place, and unless you pay for premium shipping for your supplies, they will

probably travel by a slower shipping method than the chicks (which always go *fast*). In addition, your hatchery is not going to scrutinize your order and try to guess whether or not the heat bulb you've ordered is a backup for the one you already have, whether you're getting it for a transitional period between brooder and coop, or whether you need it to arrive ASAP, before the chicks do, for your brooder. They will not automatically compare your supply order to the ship date you've chosen for your chicks, and call you to check on your intent. Instead, you are responsible for planning ahead and making sure you'll have everything you need by the time your chicks arrive.

8. **Don't wait until the last minute to order— the breeds you want may already have been reserved.** You may be surprised at how early some of the trendier breeds sell out. Even those that are more readily available may be difficult to reserve all together on the same date, so placing your reservation early is an excellent idea.

9. **Don't buy chicks or chickens as a "surprise" gift for anyone.** It sounds fun in theory—surprise, baby chicks! But pets—any pets—make terrible surprise gifts. The exception might be if you're a parent buying chickens for your family, and you'll be surprising your own children. In that case, though, expect that you'll be the one investing in the equipment needed, and that you'll be the one taking care of the birds. You

don't want to create obligations for someone else who might not be prepared for them.

10. **Don't be unprepared for delays with your order of chicks.** Hatch day issues can occur; Mother Nature works in mysterious ways. Sometimes an incubator might malfunction. Sometimes eggs just don't hatch at the rate your hatchery expected. Sometimes they'll hatch fine but a greater percentage turn out to be boys than anticipated, so there aren't enough girls to fill orders. Don't get mad at your hatchery; believe me, they'd much rather be able to fulfill all orders and make everyone happy. That's just good business. But hatcheries are quite literally in the business of counting their chicks before they've hatched. Occasional delays are just the nature of things. Some hatcheries like ours will allow you to choose whether or not you want to receive substitute breeds if there is a problem, or whether you'd prefer to wait for your breeds when they become available again. So be sure to let your preferences be known, and check your phone and e-mail frequently on the day your chicks are scheduled to hatch, so your hatchery can get in touch with you.

11. **Don't be unprepared for sexing errors.** Birds are not easy to sex. Hatcheries use experts who have trained many years, and they determine the sex of the baby chicks to about 90 percent accuracy. This means errors are expected, absolutely expected. It's also possible for someone packing your order to reach into the wrong bin, or for a bin to be labeled incorrectly. Or for a chick to jump ship into another bin. Geronimo! Things happen. You have to accept a certain amount of uncertainty, and be okay with the policies your hatchery has for addressing any problems. If you aren't, you're asking to be unhappy. Impotently unhappy, because no matter how much your hatchery wants to make things right, they don't have a magic wand.

12. **Don't be unprepared for losses.** This admonition is similar to the one above regarding errors. Losses can and do happen. Losses can occur at home with chicks you have hatched yourself, and they can even occur with a broody hen. When you are getting started in chickens, you need to be aware of the possibility that some might die. It's always heartbreaking to lose any chickens—don't

Egg-cetera!

fresh eggs

If you have a rooster in your flock, he will try to mate with all your hens, even if they are a different breed than he is—or a drastically different size.

get us wrong. But the risks and rules apply to everybody and we all have to accept them.

13. Don't count on having the chicks arrive on a specific day of the week. The hatchery has no control over how quickly the post office gets your package to you; the only thing they can do is pack the chicks safely and get them swiftly on their way to you on the date they're scheduled to be sent. Depending on the shipping method used—and how far you are from metropolitan postal hubs—it can take 1 to 3 days for your birds to arrive. The post office may be able to give you a best-guess arrival date, but just keep in mind that it's an educated guess only. So don't, for instance, schedule your last remaining vacation day off of work in advance, expecting the chicks to arrive, and then be sunk when they don't. Instead, make arrangements that are flexible.

14. Don't vaccinate without first doing your homework. Different hatcheries offer different vaccinations for chicks. Determine which ones, if any, are important to you. Also, if you're interested in complying with organic management practices, check with your local certifying agency first. Some vaccinations may prevent you from legally being able to call your eggs organic. Finally, be aware that most vaccinations are not foolproof. Vaccinations for Marek's disease, for instance, aren't effective on *every* bird,

It's important to understand that your hatchery is subject to the whims of Mother Nature and will do its very best to fulfill your order.

and even then, the vaccination doesn't prevent transmission of the disease—only expression of the worst symptoms. We still recommend this particular vaccine, but it's great if you know about these issues in advance.

15. Don't assume that the number of chickens you order is the number you'll receive. Hatcheries commonly include one or two extra chicks for free, sometimes of a breed you purchased, and sometimes not. (My Pet Chicken doesn't purposefully add extras with small orders, because we understand your coop may not have the room to house that extra chick or two, and that an extra bird may put you over your legal limit.) Also be aware that some hatcheries will allow you to place a small order for just a few birds, but will then fill out the rest of the order with "extra males for warmth." So, if you only want 5 hens, but you receive 5 hens and 20 roosters, you may not be prepared to take care of all of them, and you may not have a way to rehome the extras. If you're not purchasing from My Pet Chicken, always check to be sure they'll be sending the number of birds you ordered.

16. **Don't expect a shipping refund from the post office if your chicks are late.** The US Postal Service (USPS) does *not* guarantee overnight Express Mail delivery for live animals, as much as we would like them to. We agree: It seems a little crazy that the *most* important packages aren't subject to the same guarantees that much less sensitive packages have, but that is their policy. Visit USPS.com or inquire at your local post office for more information.

We often find that Express Mail packages are delivered in 1 day, especially to major cities and postal hubs, but, again, USPS makes no promises. Even though USPS won't refund your postage, My Pet Chicken does cover shipping costs in some circumstances, and many hatcheries do likewise. This is why it's important to read your hatchery's policies before you place an order.

ISSUES TO CONSIDER BEFORE YOU ORDER

While avoiding the common ordering mistakes just outlined will help make the process of getting new chicks much smoother, there are still other things to consider. For example, if sexing errors can happen, then how do you know how many chicks to order? Or if delays can occur, how do you know when to schedule your shipment? In this section, we address these concerns (and others) in detail and offer some guidance.

Should I Order Extra Chicks?

There are three basic questions to consider when deciding if you should order extras or not:

1. **Are there likely to be any losses or deaths?**

2. **Are there likely to be any sexing errors?**

3. **Can you keep or rehome any extras you might purchase?**

LOSSES

We get questions all the time about how many chickens are "normal" to lose from your flock. The short answer is that there isn't really an *average* number of losses you can plan for. How many may be lost over the course of bringing a flock to laying age just isn't something that will be the same for everyone, given the vastly different management styles there are. How many birds you might lose will depend in large part on

Each week, your hatchery is likely handling thousands or *tens of thousands* of hatchlings, sorting them by breed, sexing, vaccinating, packing, and shipping them–all within 24 hours of hatching.

the conditions you provide for them at your home.

If your brooder is not hot enough—or if it is too hot—you may lose some chicks. Baby chicks can drown in waterers that are too deep, and if waterers or feeders are not securely seated, chicks can even knock over a waterer or feeder onto themselves and get hurt that way. We can't predict how good your brooder conditions will be, or whether you'll have a power outage, or whether your coop and run (if you have one) will be secure from predators—or even what type of predators might be in your area. There is no way for us to evaluate all the risks there may be in your particular circumstances. Plus, even with secure, safe conditions, unexpected losses can and do happen. That said, *if you are prepared with good brooder conditions and secure sur-* *roundings, normally you will want to order the number of chickens you want to end up with.*

Losses can also happen during shipping, though they're rare. Because shipping one or two chicks alone is not safe—hatcheries have minimum quantities of chicks that they can safely ship—it will seldom be possible to send replacement chicks. So bear in mind how losses could affect your plans and your family as you're considering how many to order.

SEXING ERRORS

There is also the issue of sexing errors. There are actually 15 different "shapes" a vent sexer (someone who sorts male and female chicks by what they see inside each chick's vent on the day it is hatched) must recognize in order to tell boy-chicks from girl-chicks, and this often requires qualitative judgments, like how dull or shiny a shape is or whether it holds its shape or deflates. That's why vent sexing day-old baby chicks is an art, not a science. Vent sexing errors are rare, just as losses are, but the possibility of an error may play into your buying choices. Remember, ordering live chickens is not like ordering a shirt. If there is an error, you can't just send it back. Sometimes you just have to live with it.

Most major hatcheries offer a 90 percent sexing-accuracy guarantee. With this guarantee, your hatchery is promising that they will cover sexing mistakes of greater than a 10 percent margin of error. This means that if you order 25 hens and get 23 hens and two roosters, they've exceeded their accuracy promise. You're stuck with the roosters. The hatcheries know they can't promise there will be no errors. They're not being mean; they're not trying to cheat you. They're being honest and realistic, and you should be realistic, too. Errors happen because it's not possible to sex day-old chicks with 100 percent accuracy. Even if you experience covered errors, hatcheries cannot ship just one or two birds at a time to make up for the loss or error. In most cases you will simply be eligible for a refund. Do be sure to read your hatchery's policies for details *before* you make a purchase, as policies may affect your buying choices. You will need to take theses policies into account when planning for contingencies.

My Pet Chicken is unique in that we have a 100 percent sexing-accuracy guarantee. At this time, we're the only hatchery with such a generous guarantee. Even so, that doesn't mean there will be no errors; that simply means that all errors are covered. Just like a 100 percent live-arrival guarantee doesn't mean there will be no losses, a 100 percent sexing-accuracy guarantee doesn't mean there will be no errors. Instead, it means that *all errors are covered by our guarantees and refunded according to our policies.*

EXTRAS

You'll hear this a lot from us: **Think everything through** and be prepared for Mother Nature to throw you a curveball or two. Ordering chickens is not like ordering books from Amazon or clothing online from your favorite store. You can't always get what you want (as a wise man once said). This applies doubly to chicks and chickens. You can return a shirt that's the wrong size, but you can't return a baby chick that turns out to be the wrong sex. So, be sure to consider: If you are ordering a chicken especially as a pet for your child, what happens if this special pet is

Egg-cetera!

fresh eggs

Chicken families begin communicating with each other before the babies have even hatched. The mother hen talks to her eggs (often a purring sound), and the chicks begin peeping back to her from inside their unhatched eggs.

This 3-day-old baby chick is curious about its feed.

the chick who dies along the way or drowns in its waterer? What happens if this one pet hen turns out to be a rooster who must be rehomed?

Sometimes new chicken keepers think they would be able to manage a loss or two—as long as it's not *that* chicken. Not the one they picked out for their 6-year-old. Not the one chicken who would be laying chocolate-colored eggs. Not the one Frizzle they've been dying to get for years. Not the Speckled Sussex like Grandma used to keep—that breed was the whole reason they wanted chickens in the first place. Not. That. Chicken. So, be responsible. Provided that you have the means to keep or responsibly rehome any extras without difficulty, when you're trying to decide if you should order more chicks than you strictly need, in the following two circumstances, it's reasonable to do so: If

you absolutely must have a certain number of chickens no matter what, or if you are cherry-picking a special flock. If you look at your order and come to the realization that nearly any loss would end up being a case where you'd be upset because you can't do without *that chicken*, then you will want to think about how to manage. You might choose to order extras, just in case. Or you may decide to order an auto-sexing breed, where the sex of the chick can be determined easily by down color or pattern, such as a Red Star, Golden Buff, or Cream Legbar.

One last note on ordering extras: If you decide that ordering extras is for you, we recommend you order at least two extra chicks. Why? Because if you don't receive errors or lose chicks—as is likely—then it will be much easier for any potential adopter to introduce two (or more) birds to another flock rather than just one. If just one is introduced to a new flock, that one will be singled out as an intruder. However, if there are companions that can all be rehomed together, introducing them to a new flock is easier, since they will have each other. That way, any aggression from the new flock will be spread out a little.

Can I Handle a Delay?

If there is a problem fulfilling your chick order, you would expect to be notified by your hatchery as soon as possible. However, notification is a more complex issue for hatcheries than you might initially think. Hatcheries may become aware of shortages ahead of time if there has been an unexpected drop in laying when a breeding flock goes broody, or if a problem occurs with an incubator.

But most of the time, your hatchery won't know about a shortage until the day of hatch—after the chicks have been sexed and sorted. This means *you* won't know about it until the day of the hatch, either. That's important to understand. Hatcheries realize that this uncertainty can be problematic if you are planning to schedule time off work, but, unfortunately, there is no way for them to know about these issues beforehand.

At My Pet Chicken, we recognize that for some people, the date the baby chicks arrive is more important, while for others, getting their exact breed choices is more important. That's why we give our customers the opportunity to accept substitutions (or not) in the event of a hatch-day problem. Many hatcheries have similar systems—just let them know what your priorities are so they can handle the problem in the way you'd prefer. While they can't prevent occasional shortages, due to the nature of the business, they do genuinely want you to be satisfied with the way any problems are handled. Again—this bears repeating—make sure you're by your phone and checking e-mail on hatch day so you can make decisions quickly should there be a problem.

How Do I Handle a Substitution?

When you place an order for baby chicks, some hatcheries, like ours, will allow you to choose how you would like your order handled should there be a problem. So what option should you choose? Do you want to receive a substitute breed if the breed you have chosen is not available as predicted?

Your choice should simply depend on how you want your hatchery to handle any problems with your order. Some customers will wait many months for just the right breed, because the combination of breeds they have chosen is the most important consideration for them. Others may not care so much what breed they receive, so long as they get the shipping date they need.

Your choice may also depend on the way your hatchery handles shortages or problems. Some hatcheries may ship your order without the missing birds, while others will make automatic substitutions. Sometimes a hatchery will simply fill out your order with extra "males for warmth." Some may automatically postpone your order until a later date, while others may just cancel your order.

Here's the thing, though: If your order is

Egg-cetera!

It takes about 21 days for chicken eggs to incubate and hatch, but it can also vary a bit by breed. Some breeds may take, say, 5 hours longer—while others take several hours less to hatch. Our hatchery sets eggs so they'll all hatch (ideally) at exactly the same time.

Silkie bantams and Cochin bantams are a great choice for kids–they tend to be wonderfully friendly and docile.

like to see happen if there is a problem. You don't want to have your order delayed if you have arranged for your grandchildren to visit from across the country to see the chicks. And you don't want to receive a substitution if you have anticipated for months getting your favorite rare breed—the one that led you to order chickens in the first place. And you don't want to feel slighted if your hatchery has followed the instructions you gave them. They can't stick the chicks on a shelf and then spend several days contacting customers at their leisure. The chicks have to be shipped immediately after hatch. So consider carefully, and be clear when expressing your preferences. If your hatchery doesn't allow you to express your preferences, and if you're not okay with their automatic response to shortages, find another hatchery. The bottom line is that whether to accept substitutions or not will just depend on your personal needs, and what you want to happen in case Mother Nature does something unexpected.

postponed until the next available date, it isn't always going to be the very next week. In fact, it is seldom going to be the next week. Sometimes a new ship date can be several or more weeks away, especially with very rare breeds or with combinations of many breeds in a single order. It may be August, when you were expecting birds in April. Is that something you can stand? This is why hatcheries usually don't like to postpone orders if possible, and why they are so conservative when estimating how many chicks they expect to hatch, and how many they'll have available on a given week.

Be sure to think your decision through and let your hatchery know *exactly* what you would

What Time of Year Should I Get Chicks?

Whatever time of year you've decided to get your baby chicks, make sure you've allocated enough time to get your coop prepared. Four to 6 weeks—the time it takes for your brand-new chicks to grow into chickens old enough to be moved to the coop—goes faster than you think. Especially if you are custom building (or having a coop custom built), getting your coop ready can take longer than anticipated—and keeping nearly full-grown chickens inside your house in an area that's too small is not fun for you—or them.

Whether you get chicks in the spring or fall, keep in mind that if you are very particular about the breeds you want to receive, you may need to reserve your chicks months in advance and possibly wait months for them. It's not the end of the world. Just understand that it won't always be possible to schedule your exact ideal chicks on your exact ideal ship date. You will be happier if you are prepared to be a little flexible.

GETTING CHICKS IN THE FALL

Because there is extra effort required for summer and fall chicks, most people prefer to get chicks in the spring. Is the extra work hard? No, not particularly—not if you're prepared. However, it requires extra thought and attention from the get-go, so *be sure to prepare* if you schedule fall delivery. Sometimes you'd rather

not get your chicks in the fall, but it's the only available date—or sometimes your order might be postponed until later in the year.

Regardless of the reason, summer and fall chicks could require additional measures because it may be cold outside when it comes time for them to be moved to the coop. You don't want to move them from 70°F home temperatures directly to 20°F outside temperatures. That's a 50-degree drop, and your chicks wouldn't survive the sudden temperature change. Naturally, they'll need more transition time in winter weather.

You might consider an interim move to a garage, three-season porch, or other protected location. You could also choose to use a temporary heat source in the coop—be aware of fire hazards—and continue slowly lowering the temperature every few days until it is about equal to the temperature outside.

Chicks purchased when the weather's warm can be transitioned to their outdoor coop more quickly and easily than chicks purchased in colder months.

Next, remember that most backyard breeds don't begin laying until they're about 6 months old. This means that unless they've arrived in late winter or early in the spring, laying may be sporadic until the winter has passed.

GETTING CHICKS IN THE SPRING

Consider this: If you get chicks in the spring, the acclimation process is going to be relatively easy. By the time they're fully feathered at 5 to 6 weeks old, it will be summer and you'll be moving them from 70°F inside temperatures to mild or even hot outdoor temperatures.

On the other hand, it will be colder when they are shipping, which can increase the risk to them. As mentioned earlier, the My Pet Chicken hatchery uses long-lasting heat packs to keep the babies warm on their way—and custom amounts of ventilation and bedding, as well—but most hatcheries don't make this kind of effort.

If you want spring chicks and you are wondering how early you should schedule your reservation, you may worry: How can you predict months in advance what the weather will be when they're shipping? The answer is: You can't. But you can look at *average* temperatures for your area and schedule your ship date at a time when your *average* low temperatures do not fall below freezing, and the *average* daytime temperatures *6 weeks after your ship date* will be at least in the 60s. If you can manage that, it will help contribute to a less stressful journey for your chicks, and a smooth, worry-free transition outside when the time comes.

THE BIG DECISION: WHAT TO DO WITH OLDER HENS

There is something else you need to be prepared for before you start your hobby or order your chickens: Consider how your family will handle an aging flock of hens.

Most people who keep pet chickens allow their hens to live out their full lives, but if you have objections to keeping older chickens who lay fewer eggs, you should also be prepared to take responsibility for what to do with the hens you don't want anymore. Be sure to talk with your partner or spouse about this—you'll want to be in agreement.

Most hens live to be 3 to 5 years old, but it's common for a well-cared-for hen to live 8 to 10 years or more. We've even heard several reports of some chickens living as many as 20 years. The older they get, of course, the fewer eggs they lay—and the more seasonal their laying becomes. What are your plans as your flock ages? Before you make a decision, consider all their other valuable functions (besides being a loved member of the family): They still eat ticks, mosquitoes, and other bugs and grubs—not to mention that older hens are still fertilizer machines. Some people like to process or eat their hens when they get old. We get it—there are benefits to managing your flock this way. Presuming you eat meat, processing your own birds reduces your dependence on factory farms, allows you to have a closer connection with the way food is produced, and supports a more sustainable lifestyle. Keeping a young flock at the peak of their egg production seems

This 4-year-old Blue Cochin hen is scratching through fall leaves looking for tasty tidbits in the soil.

to make sense in a financial respect, too. In addition, the birds you have raised will have had a long and wonderful life in comparison to what they would have had in factory farms. "Meat" birds raised in factory farms are usually slaughtered at only 6 weeks old.

But if you want to live more sustainably, there are many motives for keeping old hens in the flock. If sentiment isn't reason enough, let's discuss a few additional benefits.

Cost of care is reasonable. First of all, at a backyard scale, the financial considerations of keeping older hens are relatively insignificant. If you have three, or 10, or 30 hens in your flock, you're not losing hundreds of dollars every month by feeding hens who aren't at their peak production.

Keeping older hens maintains pecking order stability. Processing birds has a drastic effect on the dynamic of your flock, because an established flock also has an established pecking order. Anyone who has ever lost a bird, though, can tell you that removing even one

bird from a flock can cause pecking order issues. This happens sometimes even in the case of a temporary loss of a hen from her regular place, for instance, when she goes broody and doesn't claim her normal place at the roost or feeder. And if losing *one* bird is problematic, removing more than one can cause bedlam, at least for a while until a new pecking order has been established.

You want a peaceful, happy flock because you want your hobby to be pleasant—who doesn't?

Introducing fewer new chickens means less work and worry. Hand in hand with the mass chaos of removing old birds from the flock goes the additional mass chaos of introducing new birds to your flock to replace the old ones. Since introductions are much easier when new babies are raised by a broody hen in your flock, this is where it's especially helpful to have older hens.

Broodies can raise new chicks. Older hens are more likely to go broody than younger hens, and to be available to raise the chicks you purchase or hatch. Experienced broodies are worth their weight in gold; chicks learn from their mother.

When you keep your older hens, your flock will have time to develop its own unique and rich culture. Older hens develop more understated communication mechanisms as their language and flock culture develop. In some cases, for example, slightly raised hackles or a dirty look (affectionately known as the Stink Eye by chickenistas) may be sufficient to communicate displeasure.

Stable flocks accept newcomers. Even nonbroody older hens often will accept chicks because they've seen it all before. Older hens are more likely to understand that new chicks aren't intruders—they're just chicks. Plus, they know how to effectively communicate with the newbies without drama, because they've communicated with chicks in the past.

Older hens provide great genes for the next generation. If you plan to breed and hatch eggs from your own chickens, keeping older hens is important, too. For instance, if you have a bird who continues to produce eggs in reasonable quantities as she gets older, this may be the very girl whose eggs you want to try to hatch. Hopefully those qualities will be passed to her offspring. Think about it: How can you make logical decisions and evaluate which bird is healthy, friendly, and productive in the long term if you get rid of your chickens as soon as they reach a year old? Or 2? Or 4? Some chickens continue to lay productively for a very long time. Although the eggs of older hens can sometimes be more difficult to hatch, these still may be the birds whose genetic material you'll want to use to increase your flock.

The bottom line is this: In many cases, keeping older hens in your flock can save you substantially down the road. It's a long-term strategy. Having a stable pecking order requires less work and provides more pleasure. It may cost less because your flock knows where to hide when predators come—and knows how to effectively teach this to everyone, even the youngest, most productive layers, who might otherwise not have figured it out on their own. And it may cost less in the long term, too, because your own flock becomes more sustainable and productive when you are making good breeding choices.

However you decide to handle the issue of aging hens, though, make sure you're in agreement with your partner or family before you begin your chicken adventure.

Egg-cetera!

fresh eggs

The age at which a rooster first crows varies, but generally speaking, he will begin crowing at about 4 to 5 months of age. There are exceptions, though: Some roosters won't begin until 9 months, while others may begin as early as 2 months!

CHAPTER 9

NURSERY to COOP

If you've decided to start your chicken-keeping adventures
with baby chicks, it's time to prepare for their arrival.
Here's what you need to know about the first 4 weeks and beyond.

You've made the decision to start with baby chicks—congratulations! We talked about appropriate brooder equipment in Chapter 6, so now you'll need to know how to make sure those babies are getting the best possible care. You'll also need to know what to expect from your hatchery and how the post office will handle your peeping package. Finally, as they grow, you'll need to know how to transition them to their outdoor coop.

HEAT: PRIORITY NUMBER ONE

As you know, newly hatched chicks need access to temperatures of 95°F their first few weeks.

Each week, the temperature in your brooder can be adjusted down by about 5 degrees until the chicks are fully acclimated and the temperature of their brooder matches the ambient temperature of the coop they'll be moving into.

If you've chosen the Brinsea EcoGlow, your chicks will self-regulate, positioning themselves as near to the heat as they care to be, as discussed in Chapter 6.

If you're using a heat lamp, it will be up to you to maintain the correct temperature. You might think it's as easy as simply putting a thermometer in the brooder. And this can work (be sure to monitor temperature at the height of the chicks—since heat rises, you may find very different temperatures depending on where in your brooder you place the thermometer). But even when you monitor temperature with a thermometer, it just doesn't always give the whole story, and it doesn't always provide your little charges with the best environment. Chicks with dark down get hotter under a bulb, and tend to stay farther away from it. Some chicks may want it warmer than 85°F in the 3rd week, while others want it cooler, plus you don't want to provide heat that the chicks can't escape.

Instead of relying solely on a thermometer, pay close attention to how your chicks behave. If they're all huddled together directly under the heat source—or if they're piling on top of each other in a corner for warmth—they're cold. And they will cry—active, distressed peeping. It's a very different sound than the contented little chirps and trills they make when they're happy. If you hear sounds of distress, lower the heat lamp or add another one. And do it quickly. Getting too cold is very bad for chicks! They can smother each other trying to stay warm in an environment that is too chilly for them.

On the other hand, if your chicks are spread out around the edges of the brooder, avoiding the heat and each other like the plague, they're too hot! When there's nowhere to escape the heat, they may begin panting, beaks open, to try to cool down. This is a warning sign. They can die from getting too hot. You don't want it to get to the point where they're panting, so make sure to remedy the situation as soon as you notice them attempting to escape from the heat bulb. Raise the heat lamp, or move it farther to one side of the brooder so your chicks will have a variety of temperatures to choose from and can relax comfortably in the temperature zone in which they are most content.

Chicks are at the right temperature when they are milling around the brooder, eating, sleeping, and quietly peeping.

Think about it: If they were nestled under mama hen, they would also come out from beneath her, experience cooler temperatures, eat, drink, scratch, explore, and return to her when they needed to warm up. If they get too hot beneath her, they can emerge again, or even just perch, relaxed on her back for a while (very cute). That's why your baby chicks should ideally have a certain amount of control over their surroundings. If they have access to only one temperature zone, it can be problematic because they may not be able to find a comfortable spot—they'll just have to swelter or shiver where they are.

A happy flock will be exploring merrily all around the brooder. You can't tell by looking at your thermometer whether or not your chicks are happy. *Observe your chicks.*

FEED

Just as puppies require special puppy food, baby chicks require chick food. Suppliers have formulated special feed complete with everything baby chicks need. It's called starter feed, and you can find it at a nearby feed store or purchase it online. My Pet Chicken offers several varieties, both traditional and organic.

Chick feed is not all that different from adult chicken food, but for various reasons, you do *not* want to feed your baby chicks regular chicken feed designed for layers. For one, chick feed is milled more finely so it's easier for the babies to eat; adult-size pellets of food can be difficult for them to swallow. Then, even chick starter feed comes in either "crumbles" or "mash." Mash is more finely ground, so if you have very small bantams, for instance, you may want to start with mash. Imagine your tiny puppy trying to eat gigantic pellets of dog food. You could crush up the feed, of course, but that's not the only difference.

Layer feed contains a lot of calcium, which hens need lots of to make strong eggshells. Baby chicks need calcium, too, of course, for strong bones. But too much calcium for chickens that are not laying—as much calcium as layer feed has—can cause skeletal problems. Layer feed also doesn't have quite enough protein for baby chicks; it just isn't nutritionally balanced for them. Would it kill your puppy to feed it adult dog food? No. But you want your animals to be as healthy as possible, and their long-term health is affected by their early development, so how you feed them is particularly important at this age.

How long should you feed your baby chicks starter feed? That's easy: Check your feed bag and find out. Different brands have different suggested feeding periods, so you will have to read the label of your particular brand of feed to see what the suggestions are, based on the formulations. Most chick feeds, medicated or not, are formulated to be fed to babies who are no more than a few months old. Some brands have a grower or developer feed that is meant to be given to young chickens after they've outgrown their chick feed, and before they're ready for layer feed.

Customers also will ask us whether they can feed their chicks scraps or worms and other bugs from the garden. Varying their diet a little is good for them, and they'll love it! But consider those extras like dessert, not the main course. Starter feeds contain everything chicks need to survive and thrive, and filling them up with too many snacks can throw off their nutritional balance. As they get older, they may be able to be more discriminate in balancing their own diet, particularly if they have a lot of range. In any confined situation, though—including in the brooder—chicks will have a tendency to simply eat what is presented to them. Like kids, right? A little snacking is okay, sure! But a lot is not okay.

You clearly *can* feed your chicks on nothing but potato chip leavings and cooked noodles, but it's not something you *should* do if you want them to be healthy and happy. Remember, your chicks and chickens will doubtless come to think of you as the Bearer of Deliciousness. They are accustomed to trusting that what you give them to eat is good. So make sure it *is* good. Even healthy foods should be eaten in balanced quantities.

A final consideration when choosing chick feed is that some feeds are medicated and some are not. Which should you get? It's up to you. Medicated feed is formulated for chicks to help them combat coccidiosis, a disease that is found just about everywhere in the environment. Most medicated starter feeds contain the medication amprolium. Amprolium in their feed does not *treat* coccidiosis—meaning that if your chicks are already sick with it, switching to a medicated feed will not cure them. They will need a different treatment plan—something like sulfamethazine sodium solution (Sulmet). (Consult your vet.) Instead, medicated feed helps healthy chicks fight off cocci oocysts while they develop their own immunity. Medicated feed is a preventive, not a cure. If your birds have been vaccinated against coccidiosis, feeding them medicated feed will nullify the coccidiosis vaccination, although it will not hurt them. But if you want to raise your chickens organically, you'll probably want to avoid medicated feeds.

It is not strictly necessary to use medicated feed at all, although some people prefer to use it as a sort of insurance policy. With a small flock and a clean brooder, you can usually use regular, unmedicated starter feed without any problems. Chicks can live and grow without medication (and have done so for time out of mind), but if you choose to use unmedicated feed, *remember that their brooder must be kept extra clean and dry, since cocci oocysts proliferate in wet environments.* That said, it is much easier to raise, say, six or 10 baby chicks with regular feed than it is to raise thousands in factory farm conditions. Conditions in factory farms often require medicated feed since the brooders are not always kept clean and dry like you would keep them at home.

Even when your brooder is clean and dry—and even if you feed medicated starter—it's not a magic force field that keeps chicks from getting sick. It's still possible for them to get coccidiosis. Sometimes an especially stressful journey can make chicks susceptible, or simply having had a hard time escaping the shell can weaken them. It could be that, although they are secure from your dog or cat, the barking or hissing—or even just a lazy cat lounging on the top of their brooder—is still scaring them. Stressors will make your chicks more vulnerable to any type of infection, so no matter which path you choose to take, be sure to keep your eyes open for signs of illness. Read more about how to recognize signs of illness in Chapter 11.

GRIT

You may have noticed that chickens don't have teeth (thus the saying "scarce as hens' teeth"). Instead of chewing their food with teeth, chickens ingest small stones and rocks. These are stored in part of their digestive system called

Egg-cetera! *fresh eggs*

Chickens go into a passive, trancelike state when they sleep at night. Consequently, if you need to catch and handle a chicken that is not very tame, do it at night after the chicken has settled down to roost.

the gizzard. As the food passes through the muscular gizzard, the stones grind around together like teeth and chew up the food. If your chicks are eating nothing but their finely ground chick starter, they won't need grit right away. That starter feed is ground finely enough that it doesn't need to be further ground up in their digestive tracts. Once the chicks begin eating other things like bugs, bits of plants, and seeds, they will need chick grit or something else like parakeet gravel added to their diet. You can either sprinkle the grit in their feed or provide it in a small cup or bowl. If your chickens have access to range, they will most likely pick up grit on their own, but it makes sense to always provide grit, just to be safe. It's inexpensive and never goes bad.

Take a peek inside–this small order of chicks just arrived from the hatchery.

WATER

Generally, we don't think it is a good idea to offer anything but plain, clean water unless your chicks are exceedingly stressed. When chicks are put into their brooder for the first time after shipping, they are understandably thirsty and will drink more and recover faster if their water is fresh and tastes good. If you have experience medicating animals, you probably know that giving them medication via drinking water is convenient but potentially problematic. If it is hot, and they drink too much, it will increase the dose they are getting, sometimes too much. On the other hand, if the medication makes the water taste bad, they may not drink very much and won't get enough medication (or water). Medicating drinking water is so inexact

that we don't recommend healthy chicks be routinely medicated, just in case. Some people suggest sugar water as a matter of course for shipped chicks, but the excess sugar can give them diarrhea: Not good!

Worse, most electrolyte formulas are designed for large commercial operations. For instance, the package instructions may direct you to dissolve the entire package in 2 gallons of water to create a "stock solution." Then, you must meter the dose at 1 ounce of the stock solution per gallon—but the stock solution keeps for only a day or two. However, we hear from people every year who didn't read the instructions carefully and poisoned their birds by offering them such a concentrated stock solution that the birds were poisoned and died very painfully. Symptoms include staggering, falling, and sometimes even foaming at the mouth.

The bottom line: Unless your babies are showing actual signs of distress, it's best to just offer plain, clean water.

ABSORBENT BEDDING

Baby chicks poop. A lot. That's because they have to eat a lot to grow as fast as they do. Don't forget that they are going from egg-size (2 ounces or so) to chicken-size (4 pounds or so) in just 2 to 3 months, and that takes a lot of energy and food. Imagine how much baby humans would eat if they grew into teenagers in just a few months. So make sure to line the floor of their housing unit with absorbent bedding, about 2 inches deep.

Let's discuss bedding in more detail. Bedding is an important part of keeping your chickens happy and healthy. In the brooder and on coop floors, good bedding will provide a soft surface for your chickens to walk on and will absorb droppings and odor. Without good, clean bedding, your coop will be a stinky mess, the same way your cat's litter box would be a disgusting mess if you didn't add any litter.

There are many bedding materials to choose from, but the best in most situations is pine wood shavings (not to be confused with pine chips). *Don't* use cedar shavings, no matter what friends or your local feed store employees tell you: The aromatic oils will irritate your chicks' lungs and make them more susceptible to respiratory problems later in life. Pine shavings can be purchased at pet or garden supply stores, and they're the best choice for most people. Some materials, such as peat moss, are just too dusty. Other materials, like straw, may be cheaper, but they can also be less absorbent or far more likely to become infested with pests (or both).

Mind you, straw has been used for a long time as bedding in coops—you can certainly do it! But because it is less absorbent, rots and molds easily, and is prone to insect infestations, you will need to use more of it and clean more frequently. Is this something you want to do? Also, if you must use straw in your coop, be sure it's straw and not hay. Hay has all the disadvantages of straw, but it's more expensive and even more prone to insect infestations than straw. Yergh. Straw is merely dried grass, but hay is dried grass with seed heads attached; it has food value. While your girls may have some fun

Saving Money on Bedding

If you want to save money on bedding, check to see if there are any unique local options. For instance, is there a wood-turning club in your area? If so, what do they do with their shavings? Lissa's father-in-law turns wood, so he gives her family his shavings rather than having them end up in a landfill. She likes them better than commercial shavings; they seem to last longer, and because they're usually thin spirals of wood, they tend to be very fluffy and absorbent. We've also heard of people who get corn-cob bedding locally, or who have access to some other unusual bedding. If you are unable to find your own special bedding source, choose hardwood, pine, or aspen shavings.

Bed your chicks on soft wood shavings, like pine or aspen (but never cedar!).

picking through fresh hay to find grass seeds, so will bugs. Don't use hay.

Also resist the urge to use newspaper no matter what others advise. It's not nearly as absorbent, and the slippery surface can lead to a permanent deformity in chicks called splayed leg, which can ultimately result in the other chickens picking on the affected bird to death. Many people also swear by paper towels, changed often, but this is fairly expensive and more time consuming than pine shavings. Plus, paper products are more likely to mold than shavings. Mold in the brooder can lead to many illnesses.

Now that the brooder is ready and you've placed your order, you're ready for your chicks. Here's what to do once they arrive.

If you are expecting to receive your chicks through the mail from a hatchery or breeder, find out what day your chicks are expected to arrive at the post office, and let your postmaster know in advance that you'll need to pick them up as soon as they arrive. Then *be ready* to receive the call. Be proactive and involved in the process. Since some chicken breeds must be reserved several months in advance, be sure to mark your calendar so you won't forget when they are coming.

Most post offices require customers to pick up packages of live baby chicks at the post office. For obvious reasons, postal workers can't leave such a package in a mailbox or on a porch unattended. In severe weather, your chicks could overheat or get rained on, or the box could be blown over in the wind. Even with good weather, they are at risk of predation from cats and dogs. That's why most post offices are required to call you to pick up your baby chicks. That's also why, when you order baby chicks from My Pet Chicken and most major hatcheries, your phone number will be printed on the mailing label, so the post office will know how to reach you when your package arrives.

Make sure to provide your chosen hatchery with the best contact number. Do you want to be called at a cell number? Your home number? An office number and extension? Clearly, if you don't receive the call from the post office, you won't be able to promptly pick up your chicks—and prompt pickup is the first responsible thing you can do to help ensure the health of your new pets.

Before your chicks even ship—in fact, as soon as you place your reservation—we recommend that you touch base with your local post office regarding their policies for handling chick deliveries. For instance, some post offices will call to confirm you'll be at home and then will deliver the chicks right to your door instead of asking you to pick them up. Others may know the approximate time that the trucks carrying such packages will arrive, so they may be able to tell you about what time to expect a call. Some post offices may call you in the wee hours of the morning or very late at night if you give them permission. Whatever the local policies, you'll need to know what to expect so you can be pre-

pared. In our experience, postal workers really bend over backward to make sure the packages are handled gently and that the chicks arrive as quickly as possible.

When you get your chicks home, be prepared for the possibility of losses. Most hatcheries, ours included, have a 100 percent live-arrival guarantee, but that doesn't mean it's impossible

Lissa's Scratchings

Stressed or weak chicks often get picked on. Sometimes, though, it's not so much that a chick is getting picked on, as it is that she's just too unsteady to deal with all the chaos and activity of the healthy chicks in a brooder. For instance, at our farm, we once had a chick hatch with a malformed leg. It was a long, rough hatch for her to begin with, and when she finally escaped the shell, she could hardly stand, even after a day in the brooder, which should have been sufficient recovery time. We realized she couldn't use one of her legs at all. My daughter immediately dubbed her Tiny Tim.

Tiny Tim seemed otherwise healthy, but she had half the mobility resources of the other chicks in the brooder. Her deformed leg prevented her from standing easily or keeping her balance once she struggled to her feet, er, foot. She would hop to the feeder and get tipped over by the other chicks as they rushed from feeder to waterer and back again. Then she struggled to get back up again; the instincts that told her how to walk told her nothing about how to manage with just one leg.

I knew that chickens could live with one leg—Teddy Roosevelt famously kept a one-legged rooster as a pet—but Tiny Tim was not doing well. After watching her progress (or lack thereof) for a day or two and seeing her get progressively weaker, I determined that she needed a separate, safe area so she could eat and drink unmolested until she was stronger and better with her balance.

Poor Tiny Tim. You might think a chick would be relieved to be able to eat and drink without being knocked over, but chickens are flock animals and won't be happy unless they are part of the group. So, I took a bit of chicken wire and put it in one corner of the brooder, simply creating a separate space for her where she could still see her siblings. However, chicks don't normally want to be separated for any reason, even when it's for their own good, and that was certainly true of Tiny Tim. When she was convalescing in her safe area, she cried piteously almost the entire time. It was heartbreaking, but I wasn't willing to let her slowly starve to death.

Still, because of the way I set up her space, she was so close to the others that she could see them and interact with them. She often even slept snuggled against her siblings, right next to the wire. She could access for chicks to die along the way. The live-arrival guarantee means you will be compensated for losses, not that hatcheries can magically make chicks immune from death. When it comes to major hatcheries shipping baby chicks, the risk of loss is small; in fact, it tends to be less than the risk of losing baby chicks hatched in a home incubator. However, even broody hens lose chicks, so you'll want to be prepared for the possibility that you might lose some.

Why do some chicks die during transit, while others in the same package arrive happy and healthy? The fact is that some chicks are simply less hardy than others when they hatch, for various reasons. Factors like eggshell porosity and thickness can affect how difficult it is for a chick

the same feeder and waterer that her sisters could—she used the same heat lamp—but they were unable to knock her over, and she grew strong enough to stand more steadily, maintain her balance, and hop where she wanted to go.

This is the ideal recovery situation—and be prepared that even this will make a chick unhappy. You might equate it to cleaning your child's scraped knee: Rinsing the wound with cool water and mild soap might sting, but it's necessary to try to prevent infection.

What you may experience in your own brooder could vary. More commonly, for example, a chick can get injured. And a little scrape or wound can turn into a serious, life-threatening injury when it gets pecked and prodded at repeatedly by other chicks. Unfortunately, the instinct chicks have that helps them know how to eat on their own rather than wait to be fed—pecking at anything interesting—is the same instinct that can cause them to hurt an injured chick, by pecking at her wound. When that happens, you will need to separate the injured chick until she has recovered. If you can keep her close to the rest of the flock, she will be more comforted by having company, and she'll also have less trouble reintegrating into the flock when she's better.

With Tiny Tim, after 3 days, I removed the separating wire in the brooder, and she rejoined the flock. Her leg was still deformed, so she still had to get by on just one. And she still occasionally got knocked over; that would never change. But I'd given her the time she needed to figure out how to manage with her disability. When she did get knocked over, she was now fit enough to just get back up and continue on her way. Tiny Tim is still doing well today.

When creating a space for your chicks, remember that all of them will need appropriate warmth, food, and water, and they generally feel better when they can see each other. And, keep in mind, if you think one of them is ill (not just stressed or weak from shipping), separation will help, but you'll need to get her to a vet for a firm diagnosis and treatment options—because if she's ill with something contagious, you'll want to know right away so you can protect the rest of your flock.

to escape the shell. An especially harrowing hatch can mean the journey will be stressful, too. If your chick had a rough hatch and then a rough journey, she may not make it.

Many hatcheries prepare for the possibility of losses by including an extra bird or two for free, but there are a few reasons this isn't always the best idea. At My Pet Chicken, for instance, our customers are generally looking for small backyard flocks; they may have ordered four chicks because local regulations say that they're only allowed to have four chickens, no more. We don't add chicks to small orders of fewer than 25 chicks, in an effort to be sensitive to cases like that.

STEPS TO TAKE WHEN YOUR CHICKS ARRIVE

We discussed brooder equipment in Chapter 6, so when your chicks arrive, you should be ready to go. Make sure there is plenty of space, good litter, appropriate warmth (95°F), fresh finely milled chick food, and clean water. What else do you need to do to provide proper care?

1. **Count to see if you've received all the chicks on your reservation.** If you think you haven't, check your original order. (You may be surprised by how often memory can fool you.) If you haven't received all your chicks, contact your hatchery.

2. **While you're counting your chicks, also try to verify that you received the correct breeds that you ordered, since packing errors can sometimes occur.** In some cases, errors may not be immediately obvious: New Hampshire Reds, Rhode Island Reds, and Red Sex Link hybrids like Red Stars and Golden Buffs are very similar in appearance as baby chicks. In other cases, mistakes will be hard to miss:

TRACI'S SQUAWKINGS

Handling chickens is an art, and practice makes perfect! The key is finding the balance between being gentle and letting them know that no matter how much they wriggle or squirm, they're not getting away.

First, put your dominant hand (the hand you write with) on the middle of the chicken's back. If you're new to chickens, it's helpful to secure the wings as much as possible with your thumb and forefinger. (Pros don't need to do this.) Your other hand will need to take the chicken's legs out of the equation. Secure one leg between your thumb and forefinger, and the other between the forefinger and middle finger of the same hand. Then lift the bird, supporting the lower portion of her body with the heel of your hand and wrist. Your dominant hand should still be on her back. Once you've got her up, holding her close to your body will prevent further wriggling. And, again, as you get better at this, you won't need a hand on the chicken's back.

Bearded, crested Buff Laced Polish chicks will look very different from Black Australorp chicks without any fancy feathering. If you think there has been an error, first look again at your original reservation to verify what you should have received. If there has been an error, contact your hatchery. Most hatcheries will refund you for such problems, but only if reported within 48 hours of receiving your chicks.

3. **Next, check each chick for pasting.** Pasting occurs when droppings have dried around your chick's vent, sealing it shut. Obviously, if your chick is unable to have a bowel movement, she'll die in pretty short order, so the pasting needs to be removed immediately. Keep in mind that having some droppings on your chick's down, while unsightly, is not pasting, and the chick will eventually preen it out— you needn't intervene in that case. It's only when the chick's vent is sealed shut that there is an immediate danger. To remove stubborn pasting, you can use a warm, wet paper towel to soak the droppings off. **Be sure that the chick doesn't get too cold** during this process; chicks that get chilled are more likely to have loose movements, which are more likely to get stuck in down and cause problems. Clean her quickly and in a warm environment. Whether she gets too cold or not, she will likely complain from all the handling and stress. No one wants to make a baby chick cry, but this is necessary; she will die if she can't move her bowels. When the material is removed and the vent is clear and open, dry the chick off with a blow dryer set to warm (not hot), so she doesn't get chilled before you return her to your brooder. Be sure not to overheat her, either. She must be dry before she is returned to the brooder; pasting can become a vicious cycle if the chick gets too cold during or after the removal process. In especially bad cases of pasting, you may have to dunk the chick's rear in warm water before the dried matter will loosen up enough to remove it. For chicks who've had the pasting issue, keep checking for a week or so, because the pasting can return for a while until their digestive systems have stabilized and they've recovered from whatever caused the pasting to begin with.

Egg-cetera!

fresh eggs

Coincidentally, most chickens with white earlobes lay white eggs, while most with red earlobes lay brown eggs. However, this isn't a rule. For instance, blue and green egg layers usually have red earlobes, too. And Penedesencas and Empordanesas have white earlobes, but lay dark brown eggs.

A little dried poo on a chick's fluff isn't a problem. You only need to worry if the vent opening is fully sealed shut.

4. **To encourage your new birds to drink, dip their beaks in water as you check them for pasting.** What we find to be more helpful—and less stressful than forcibly pressing their beaks into a water dish—is to place a few sanitized stones or marbles in their waterer (boil stones in a covered pot for 3 minutes). This extra visual stimulus encourages them to peck at the water and discover how to drink.

5. **Be aware that some baby chicks will arrive at your home with an intact, attached umbilical cord—don't mistake this for pasting.** On your chick, an umbilical cord may look like a very thin black string, attached to their rear, a little beneath the vent. You can dab a little iodine

on it if you fear it might be getting infected; that can sometimes speed drying. In most cases, you'll just want to leave it alone. Sometimes there will be no cord or string, just a scab over the umbilicus. Again, don't mistake this for pasting, and for goodness' sake, don't remove it! Just leave it alone, or, if you must, dab with a little iodine. Removing the scab (or even just pulling on the umbilical cord) can kill your bird painfully and quickly.

6. **If you experience losses, bury the dead birds as you would any other pet.** Also, call the hatchery right away to let them know if a bird dies. They'll usually either give you a partial refund or a discount on your next order so long as you notify them within 24 to 48 hours of arrival. (My Pet Chicken's policy is 48 hours.)

CHECKING ON YOUR CHICKS

Although chick care is easy—they don't need much!—you will still need to check the brooder several times a day through the first few weeks. One reason is that as the chicks scratch around in the bedding developing their foraging skills, they'll kick the litter right into the feeders and waterers, so you'll have to be on top of things to keep it all clean for them. You don't want the babies to be drinking poopy water. Water dirtied with litter (and bedding damp from spilled water) creates a risk of coccidiosis. As your chicks get older—and taller—you can raise the height of the waterers and feeders so that they get dirtied less often, but when they're babies, the food and water will have to be close to the floor and thus will get dirty frequently and easily. Baby chicks just require more attention from you than older birds will.

Another reason you'll need to check on your chicks often is simply so you can observe their conduct and look out for signs of trouble. You'll have to determine whether they are acting too hot or too cold, or whether one of your chicks is inactive, indicating the possibility of sickness like coccidiosis, one of the most common killers of young baby chicks. Remember, your chicks can't tell you they're sick; you will have to watch them and notice there's a problem. If you don't observe them, you may never know they're ill until it's too late, because—as with other pets—the only clues you will get about their health at this age require direct observation. Worse, because prey animals like chickens have an instinct to hide signs of illness, your chicks are not going to make it easy for you to notice there's an issue. Keep your eyes open.

Change the bedding in their brooder frequently to keep their area clean and fresh smelling. How often you'll need to do it will depend on the size of your brooder and the number of chicks you have in it, as well as their size and age. Usually, brooder bedding will need to be refreshed about once a week. As the chicks get close to the age at which they'll be moved outside (at 4 to 6 weeks, when they are fully feathered and acclimated to outside temperatures), you'll probably need to change bedding more often, since they'll be large enough to produce more droppings. See "Transitioning to the Coop," on page 135, for guidelines on acclimating chicks to outdoor conditions.

If you are noticing an ammonia smell, you've waited far too long to change the bedding. Remember, your baby chicks are just a few inches from the bedding; anything you are smelling from your vantage point will be far more concentrated at their level. If you smell ammonia, your chicks will have been suffering for some time. Change the bedding immediately, and make sure to change it more often in the future. It's not hard to do; plus, the used bedding makes fabulous compost for your garden. Mix it into your compost pile and age it until it's turned into rich, black earth. (Chicken manure is too "hot" to put directly into your garden without aging first.) You'll be amazed at the size and quantity of your squash and tomatoes.

DEALING WITH WEAK CHICKS

You may have a chick that is weak from a rough hatch, or you may have one arrive weak from a rough journey or other reasons. You can tell a chick is weak if she is inactive, isn't eating or drinking, stands with wings drooping down, or if her feathers or down are ruffled out. It may take a while of watching to identify if your chick is weak or just sleeping. Remember, they're not all on the same schedule, so some may be napping while others are eating or drinking, and vice versa. If you watch carefully and come to the conclusion that you have a chick that is genuinely weak, what do you do?

First, make absolutely certain that any weak chicks are warm enough, because weak chicks may not be able to find their way to the warm, 95°F side of your brooder. Next, check your weak chicks for pasting. You will have checked for pasting when they first arrived, but sometimes pasting can occur after arrival, so keep an eye out.

If the chicks are warm enough and the problem isn't pasting, you can try some other things. Dribbling a few drops of sugar water alongside their beaks can sometimes help give them enough energy to eat and drink on their own in

Lissa's Scratchings

The first time I ever met a flock of chickens in person, I was instantly enchanted. I was visiting the farm of a distant cousin; as I pulled into the drive of her little farmhouse, I came upon a flock of bantam Easter Eggers, scattered around pecking. At that time, I had no idea there was such a thing. I thought all chickens looked, well, like chickens. And I had no idea there were bantam (or miniature) chickens. But these Easter Eggers didn't look like chickens at all. They weren't fat, for one thing. They didn't have big, floppy combs and long hanging wattles. They looked like tiny little hawks or jays, all sleek and sophisticated. And they came in different colors: black and blue and gray and red; white and gold and brown and buff—and they were patterned with lacing and speckles and spots, with bands of color on their wings, and lustrous feathers in their tails. Their faces were wreathed in fluff. Plus, they were graceful and fast, without an awkward waddle to be seen.

I was breathless.

They were calm, not in the least disturbed that I was walking among them. They had been raised with kindness and care, and, in their world, humans were not to be feared.

My cousin noticed my interest immediately, and after the traditional West Virginia welcome of a genuine, friendly smile and a warm hug, she gave me a few handfuls of scratch to toss around for her little birds. It didn't matter that they didn't know me. They came running; they were so excited. I was excited, too. And I couldn't believe it. Chickens were not what I had expected—not at all! Don't get me wrong; I grew up wanting to move to the country and get chickens, before it was common to keep a flock in town. So I had certainly always wanted chickens, and I expected to like them. But it had never occurred to me that real chickens would come in such an array of colors and shapes, that they would be so elegant, or that they would actually be friendly. Extraordinarily friendly.

My cousin's little feathered beauties were amazing. They came running fast to eat the scratch I scattered, and some even ate out of my hand. One actually flew up to my shoulder, promptly sat down, and made herself comfortable, alternately preening a wing and keeping a watchful eye over the rest of the flock, with a haughty air that I found endearing.

earnest. This is not good long term, of course, as sugar has no nutrition to speak of. Plus, too much sugar can give them diarrhea, which could *cause* pasting. However, the hydration and the burst of energy from the sugar can sometimes help them get over the hump so they'll begin eating and drinking on their own.

If they are eating on their own, but are just a little unsteady, you can also try warmed, plain yogurt mixed with their food. Believe it or not, finely chopped, hard-cooked eggs (or scrambled eggs) can also help, and are high in the nutrition they need. Again, these are not good long-term foods, but they can help a stressed chick.

My cousin took me on a tour of her coop and the rest of her farm, answering all my questions and sharing stories in the lilting central West Virginia dialect that I love so much. Finally, we sat down for some "visiting time" and a huge homemade lunch. There's a reason why West Virginia is referred to as Almost Heaven— many reasons. But one is that having people over for a delicious, homemade meal is practically the state pastime. You know you're visiting a "true" West Virginian when you're lucky enough that your meal includes local fare like venison, wild blackberries, ramps, field greens, morels, or produce from the garden or farm. The meal she made—with eggs from her little flock—was amazing.

She packed me up two dozen tiny blue and green bantam eggs from her hens and a few other gifts for the road. It was the eggs I was most captivated by, though. I gawked at the colors. As if the chickens themselves weren't charming enough, they laid colored eggs?! There were green eggs, turquoise eggs, olive eggs, and sky blue ones. There were sage-colored eggs speckled with brown, and a couple that were creamy pink. Surely they were laid by fairies or leprechauns—not chickens. They were certainly too pretty to crack and eat.

And to top off the visit, as I stepped onto the porch to go, there was a rush of air, the noise of wings, and the sound of thrumming feet. The chickens came stampeding out to meet us, and they even followed me to the car, making excited little clucks and chirps. I was again taken aback by how cheery the little birds were, and how friendly. "Why are they following me?" I asked, reaching down to pet one particularly friendly hen, the one who had perched on my shoulder and hitched a ride while I was touring the farm.

"Cause you fed 'em, and you was nice to 'em," my cousin said simply. "Them chickens knows ye. Chickens ain't stupid."

Chickens ain't stupid. She was right. She was right then, and she's right now. Even when you move your little babies out to the coop, your chickens will know you. Don't worry that they'll suddenly stop being friendly; they won't stop. It may take them a while to get used to changes in their environment, but they remember where their food comes from; they'll know where a warm lap is, or from whom they can hitch a shoulder ride.

Cause chickens ain't stupid.

Finally, please remember that these techniques can help chicks that just need a little TLC to recover from a stressful hatch or journey. If you have acquired chicks from a non-NPIP (National Poultry Improvement Plan) source and think one might be ill with something else, you'll need to get her to a vet for a firm diagnosis and treatment options.

CHILDREN AND CHICKS

Anytime your children have handled the chicks, make sure they know to thoroughly wash their hands. Remember, your baby chicks have been walking around in litter all day, getting poo on their toes. When the kids hold the chicks, they are getting those litter-y toes all over their hands. Young children in particular are prone to putting their hands in their mouths without thinking, and if they have chicken litter on their hands, that's just not a good idea. It's a terrible idea, in fact.

Of course, chicken manure isn't really any dirtier than other animal droppings, but germy chick feet are a risk you may not properly think about when looking at the little puffs of adorable down. You want to snuggle chicks, toes and all, which is usually pretty different from the way you feel about the pooper-scooper you use to manage waste from other pets. Since you probably wouldn't consider having your child clean the cat's litter box or scoop dog poop without washing her hands afterward, use the same common sense when dealing with chicken droppings: Wash hands. We also recommend a hand sanitizer; to make it easy, keep one right near the brooder as a reminder.

Secondly, especially with young kids, you'll need to make rules to protect the chicks from overeager snugglers. Tell your kids in advance what you expect from them, and the consequences for not following instructions. Be prepared to supervise them closely. And leave chicks that have suffered from pasting alone to recover (if they are very weak or are getting picked on, you may need to separate them from the flock for a time); stress from handling can exacerbate the pasting problem or cause a recurrence.

If your children aren't careful with the chicks, or if your chicks show signs of distress, be firm and return the chicks to their brooder. Having your chicks socialize with their human family is good, but use common sense. Chicks need a touch firm enough to keep them from jumping out of a hand (falling too far and hurting themselves), but gentle enough so that they aren't being squeezed too hard. They're delicate. Be prepared to supervise closely so you can teach your kids how to find this happy balance. You don't want your child's first memory of chickens to be the memory of how they accidentally squeezed or dropped one to its death.

ACCOMMODATING YOUR GROWING CHICKS

You remember from Chapter 6 that baby chicks need appropriate warmth in their brooder. When they have just hatched, they will need a warm, 95°F side of the brooder, and that temperature can be reduced by about 5 degrees per week until they are 5 to 6 weeks old and have acclimated to ambient temperatures. By the 4th week, for example, the temperature on the warm side of the brooder should be about 80°F. But remember that this "5-degree decree" is just a rule of thumb. Sometimes chicks may want it hotter or cooler than the suggested temperature. Be sure to adjust the heat based also on your chicks' behaviors, as discussed in Chapter 6. Tune in to the clues they're giving you about their comfort level, and be sure to give them the conditions they need as they grow and their requirements change.

Speaking of growing, after a week or two of having your chicks, you'll start to understand why we recommended a minimum of 2 square feet of brooder space per bird. It seems like a lot for a tiny little puffball no bigger than an egg—but it's not much for larger birds, and your babies will grow quickly. Your flock will get restless and frustrated in small spaces. With less than 2 square feet per bird, your chicks may experience a "behavioral sink," and begin picking at each other, even hurting one another. Overcrowding triggers a stress response, so to keep them happy and healthy, you'll always need to provide plenty of space and keep them from getting too bored.

One good way to combat boredom in a small brooder is to give the chicks the opportunity to explore the great outdoors. Once they are a few weeks old, you may decide to take them out for brief excursions. If you have a fenced run, you can put them there. Or, you can improvise a "playpen" from a length of chicken wire or hardware cloth joined to itself in a circle. They'll probably be unsure and nervous at first in the new environment, but they'll get over that quickly and learn to love it. Being able to taste grass and bugs—and get sunshine and fresh air—is good for them. Be sure it's warm enough for a brief time away from the brooder. Also make sure it's not too breezy or windy during their time outside. Remember that baby chicks should be kept away from drafts until they're fully feathered.

Give them access to sun and shade, feed and water, and be sure also that you have a secure pen to keep them safe. The chicks are vulnerable and they are surprisingly good flyers at that age, too, so make sure to have their playpen enclosed at the top, as well as the sides. And make sure you stay outside with them to keep neighborhood cats and other would-be predators away.

If you're worried about taking your chicks outside at such a young age, remember that if

Egg-cetera!
fresh eggs

Chickens can recognize up to 100 faces!

they were being raised by a broody hen, she'd have them outside from the start. Keeping that in mind, you may wonder why you have to wait a few weeks—or any time at all—to take them out yourself. The answer is that you don't, not really. But the reason it's usually wise to wait is that a broody hen is more suited to keeping her babies protected from the weather than you are. When the chicks are with their mother hen, they can snuggle into her feathers, protected from drafts and chill, at any time. This just won't be the case with *you* as the mother. By 2 to 3 weeks old, though, baby chicks would be spending more time outside of mom's feathers, exploring the world. That's when it will be safer for you to give them some time outside.

You'll know when hatchlings are comfortable with being handled! They'll relax and even fall asleep right in your hand.

HAND-TAMING YOUR NEW PETS

In general, chicks are friendly and tamable by the same sort of methods you would use with a dog or cat. Presuming they are healthy and receiving proper care, they will be tempted by treats (given in moderation) of wild birdseed, mealworms, grit, and small sunflower seeds. If they learn to associate you with food, they will come running to you when they hear you, rather than running away from you.

Even so, chickens do not usually like the same kind of handling a dog or a cat might like, although there are many pet chickens who love to be petted and scratched. In addition, the breed of chicken, and even the personality of the individual chicken, will have a lot to do with how friendly she will be. Some chicken breeds are flighty and startle easily; they may avoid human contact out of instinct. Other breeds are naturally friendly and calm. Choose your breeds wisely; choose based on the qualities you want your flock to have. Don't get nervous birds and then profess to be surprised that they're not friendly. Still, any chicken can be tamed with enough patience and effort. How much do you want to expend?

When it comes to hand-taming chickens (getting them used to your hand, and possibly even willing to eat from it), treats are helpful. It takes kindness and patience to make any animal friendly to you, and chickens are no exception.

When you begin with baby chicks—and it's easiest to begin when they are young (as soon as you get them, whether you hatch at home or purchase them)—here's a technique you can use to teach them that it's safe to eat from your hands.

We recommend that you choose a good-for-them treat such as ground mealworms or oats. Spread a few treats on a paper towel or wooden board laid on the brooder floor and wait until the chicks begin to eat. Chicks generally prefer to forage for food on the ground—it's their natural instinct—so in most cases, you can even use chick starter rather than special treats. Sometimes they won't recognize what a treat is, at first, so you will have to start slow and be patient. It usually doesn't take long for them to catch on.

Once the chicks begin eating reliably and calmly from the board, put some of the treats or food in your hand and lower it into the brooder. Anything coming from above will likely startle the chicks, but they should get over it quickly and venture back to eat. Soon, the boldest chicks will approach and eat the treats right from your palm. Just stay calm; keep your hand in there, unmoving, while they eat. So long as your chicks stay warm and do not get stressed, frequent gentle handling will not hurt them, and will help to make them tamer when they get older. Eventually, even the shiest chicks will follow their sisters' lead and eat from your hand.

Sooner or later, you can stop putting the board into the brooder at all; just use your hand to distribute the goodies. The chicks will jump up on your hand to eat; they may even scratch at the feed or oats in your hand (which is absolutely adorable). Then they may perch on your hand and want to go to sleep from this new, cool vantage point. If you are using mealworms, be sure to have the family handy to watch. (Mealworms are sold for feeding birds and as fish bait.) We won't spoil anything by telling you what will happen—but it will likely be the funniest, most entertaining thing you have seen in a while. Again, if your chicks have never seen mealworms before, it may take them a while to figure out what they are, and that they're good to eat. But once they've figured that out, believe me: Watching your chicks with some mealworms will be far more entertaining than TV.

Important: Make sure to provide grit to baby chicks that are eating treats, so they will be able to digest them properly. (Chick starter is finely milled enough that no additional grit is needed.) Make sure, too, that the bulk of your babies' diet comes from starter, and that they are not getting so many treats that it upsets their nutritional balance.

TRANSITIONING TO THE COOP

By 4 to 6 weeks of age, your baby chicks will be babies no more. No matter how many times we see it, it always amazes us how fast chicks grow—and how quickly they'll come to look like chickens rather than chicks. By the time they're ready to move to the big-girl coop, you will be more than ready to move them, too. As much as you love them, you'll have come to realize that they need more space—and that inside your house is not the best place for a flock of chickens. Your young chickens will need access to an outdoor run, to places where they can scratch and dig and dust bathe. They'll need space to be chickens.

The transition to the coop should be relatively trouble free if you use common sense. **The first thing to remember is simply to be prepared with a coop when the time comes.** If you are ordering rather than building, be sure to take into account delivery time. Many high-end coops are custom built, and may have a lead time of 6 weeks or more between ordering and delivery. Other coops, especially those that aren't custom built, may take just a week or two to be delivered—presuming they're in stock. If you're building it, be sure to give yourself enough time to get finished. You really don't want to be ready to move your large, 6-week-old chickens into their coop only to realize that it will be 3 more weeks (or longer) before their coop is ready.

The second thing to keep in mind is that this transition is also a process of teaching your chickens where home is. Have you ever heard the saying that "chickens always come home to roost"? Well, it's true. Once your chickens have imprinted that the coop is their home, they'll always return to the coop at dusk and put themselves to bed. That's good news, because "herding" chickens isn't exactly easy! So don't rush this process. Leave them inside the coop—with no access to their outdoor run—for at least 2 days (or more, if possible) after you have transferred them from the brooder. (Make sure they have access to food and water in the coop, of course.) This will help your birds to regard the coop as home. They will get used to sleeping there and will feel safe returning there at night.

The last thing to remember is that it's best to make the transition during mild

Once your chickens learn where "home" is, you can let them out to roam.

weather. Sudden temperature changes are a real danger to your birds. You will need to make sure their transition outside doesn't shock them. For instance, you do not want to keep your birds shut inside a hot coop in the dead heat of summer when they may get overheated.

Alternately, if it is icy cold outside when they have reached the age of transition, you'll need to make a few extra preparations. Although chickens bear cold weather quite well, they will need to slowly acclimate themselves to the new conditions. So, if it's significantly colder outside than it is in their brooder, you may want to provide an additional heat source in the coop and slowly lower the temperature every few days until it is about equal to your local temperatures (making sure you don't create a fire hazard). You could also consider moving them temporarily to a garage, three-season porch, or other protected place. We don't recommend providing a heated coop long term, but for short-term use in the transition to the big-girl coop, it is sometimes essential.

A Little Reassurance

We get a lot of anxiety-fueled questions about the move from brooder to coop from people whose chickens are old enough to make the transition. Somehow it seems cruel to them to move their birds (which they consider beloved pets) to the coop. They ask questions like, "Will my birds be safe?" or "Will they be upset?"

The truth is that any change, much less a move, can be unsettling for the chickens; chickens just don't much care for any change in their environment. A bag of grit newly placed in the corner of the coop can be regarded as an enemy for days, first avoided and then roundly scolded by any passing hen—because your chickens are not always able to immediately assess whether a new thing is a dangerous thing. A new nest might stay abandoned for months. Young roosters that are experiencing their first time outside may screech predator warnings about every flying sparrow, falling leaf, or flitting butterfly.

But they'll learn.

So, when your chickens are first making the transition to the coop, just keep in mind that they're going to have to learn where home is all over again. Home was the brooder; now it's somewhere else entirely, and the move may seem like a terrible upheaval, even if it's just a matter of moving to the yard from the house. As far as they're concerned, their home has suddenly become inaccessible. The space in the coop will be different from their brooder, and they won't have the sense of security they had when they were familiar with the routine of their old home. In addition, if you have raised your chicks with a heat lamp rather than a unit that doesn't emit light (like the Brinsea Eco-Glow), they will probably also have to get used to nighttime darkness, so they may initially be afraid of the dark. (When that's the case, a little battery-operated tap light may help ease the transition.)

It will all pass, don't worry.

We also hear questions like, "Will they forget me if they don't live in the house anymore?" or "Will they cease being tame and friendly?" The answer is "No." They won't suddenly stop being friendly just because they're moving to the coop. They're going to be scared of the new things, but not the things they trust—things like you. If they have learned that you are trustworthy, that won't change just because they've moved outside, although the stress of the change can make them a little jumpy for a while. It will be okay though.

CHAPTER

10

CARE of ADULT CHICKENS

One of the things that make
keeping pet chickens so wonderful is that
caring for them is simple.

There are only a few things you'll need to do. Daily, you'll want to **gather eggs**—and heck, we're not sure if gathering eggs qualifies as a chore or a perk, because that's one of the fun parts. You may find yourself going out to the coop in excited anticipation several times a day when your birds begin to lay, just for the pleasure of it. The eagerness for laying to begin can be so intense that you may find yourself posting photos of your first egg on Facebook or e-mailing them to friends and family, as if they're pictures of your new grandchild. But, you don't *have to* gather eggs that often. Once a day is plenty.

Depending on the type of coop and the flock-management method you've chosen, another daily chore you may have is to **open the coop door in the morning and close it again at night** to give your chickens time to range. Or you may not have this duty at all. See Chapter 6 for more details about how your coop choices will affect this.

Weekly, or thereabouts, you'll also need to **clean and fill their feeders and waterers**. Exactly how often you have to refill food and water will be determined by how large your feeders and waterers are, how many birds are using them, whether you're using a hose-fed waterer, and so on. Most people choose feeders and waterers large enough to last their flock for a few days, up to as much as a week or more. Larger sizes are heavier and less likely to be overturned accidentally by eager feeders; on the other hand, they also take up more space, which can be challenging in a small coop, and they're *heavy* to lug around when filled. You might consider buying a used child's wagon to make your life easier. Unless you have a particularly mischievous bird, you'll probably only need to clean when refilling.

Somewhat less frequently, you will have to **change the bedding** in your coop, just like you would change the bedding in a hamster cage. You want the coop to be clean and fresh, not only because—of course—a clean and fresh coop is nicer for you, but also because if you don't clean your coop frequently enough, *the buildup of odor and ammonia can be damaging to your chickens' respiratory systems* and will make them vulnerable to illness. If *you* can smell it, you've waited too long to change the bedding.

How often will you have to change it? Again, it depends—on how many birds you have and how much space you've given them. If you have lots of birds together in a small coop, with just the minimum space allowance, you'll probably have to clean once every 2 or 3 weeks. But if your birds have lots of space in the coop and/or extensive pasture access (which would include chicken-tractor management styles), you may have to change bedding only once a season or so, perhaps twice in winter if they spend more time indoors.

There's another option, though. You may also want to consider the *deep-litter method* for your coop. This will reduce the amount of coop cleaning you do to only once or twice annually. To use the deep-litter method, simply start out with an abundant layer of bedding on the floor of the coop, 4 or 5 inches thick. (Pine shavings work best because they are absorbent and don't decompose quickly.) Monitor the bedding situation in the coop about once a week, and if you see moist clumps of manure in the bedding, break them up with a spade, rake, or shovel, and turn the bedding over, mixing it into the non-clumpy bedding. Add a little fresh bedding on top as needed to keep the litter in the coop dry—but you needn't remove the old bedding first with the deep-litter method. Normally, your chickens will keep the bedding broken up for you with scratching, but the area under the roost can get ignored. If the level of bedding in the coop gets too thick and you can't open the coop door—or if adding bedding to the litter doesn't dry it out adequately—that's when you can remove the old bedding to the compost pile and start again.

Once or twice a year, whether you choose to use the deep-litter method or not, **do a complete and thorough coop sanitization**. For an inexpensive coop cleaner/sanitizer, you

Egg-cetera!

Chickens are omnivores. They naturally eat plants as well as insects, grubs, and even small animals, including snakes, frogs, and rodents. Buckeye chickens are known for being particularly good mousers.

might try a concoction of 1 part liquid household chlorine bleach, 1 part ecological dish soap, and 10 parts water. Star-San is also a good sanitizer for the coop.

Purchase a scrub brush dedicated to your coop, and scrub roosts, nests, food and water containers, walls, and anything else you might have in the coop, such as feed storage containers and treat holders. Let everything dry thoroughly before replacing the bedding. If you are using the deep-litter method, be sure to return a little of the old bedding to the coop; beneficial microbes build up in the litter over time and partially compost the litter while it's in the coop, so leaving a little is thought to help inoculate the next load of fresh bedding.

FEEDING GUIDELINES

Many people want to know exactly how much their chickens will eat, so they'll know how much feeding them will cost, and also how often they'll have to refill feeders and waterers. A basic estimate is that chickens will eat $1/4$ pound per bird per day. *However, this estimate is based on the conditions commercial layers face in factory farms, and is not necessarily an accurate estimate for backyard chicken keeping.*

This estimate is for high-production, economical producers of eggs in controlled conditions year-round, who may be highly stressed and who certainly have no access to pasture. Such estimates will not be accurate for bantam or heirloom breeds of chickens kept in backyard conditions. The differences are many.

Pet chickens:

• Have varying egg production rates, based on breed and season

• Come in varying sizes from teeny Seramas to giant Brahmas

• Often get treats like fruit and veggies from loving caretakers

• Won't be slaughtered as they get older and their laying habits change

• Will likely have grass and bugs to supplement their feed

• May have icy, dark coop conditions in the winter as well as very hot, bright summers (or not)

It's definitely not comparing apples to apples. Backyard chickens may eat more—or they may eat less—than their commercial counterparts.

How much feed they require and how much you spend will depend on a variety of factors: How many chickens you have, what brand of feed you get, what the prices are from your chosen supplier, whether you are buying starter or grower or layer feed, whether your chickens can supplement their diet on pasture, whether it is winter or summer, how long they free range during the day, what their range is like, what breed(s) of chicken you have, among others. Even the type of feeder you buy, whether it is hung or mounted at the proper height for the size of your birds (feeders and waterers should be at their shoulder height), and whether you buy pellets or mash will also affect how much feed is wasted or "beaked out." In other words, this is not an easy question to answer accurately for backyarders. (Finally, don't forget to consider that if you haven't taken measures to keep rodents out of your coop, you could be feeding them, too!)

So ¼ pound per bird per day can serve as a vague guideline, but we recommend you determine where you will be purchasing food for your flock and check out their feed prices, just as a starting point. If you are interested in maximizing your feed savings for your flock, you will want to take into account how much your breed eats to produce each egg; some breeds are "economical eaters," and convert food into eggs more efficiently. You can keep track of how much a bird eats and how much she lays to calculate the conversion rate.

There are a few chicken breeds that are known as economical eaters (Ancona, Andalusian, Campine, Fayoumi, Hamburg, Houdan, Leghorn, and Welsummer). Many of these breeds are known as flighty rather than friendly, though, and most are not good winter layers. Some are known to go broody, which means they will stop laying eggs occasionally because they will want to hatch some. So, the best breed on this list for backyard pet purposes might be the Welsummer because it is described as friendly, is a good layer of dark brown eggs, and also cold hardy enough to continue laying in cold weather. (The Welsummer is not known as a particularly broody breed, although individual hens may vary.)

PROVIDING PROPER NUTRITION

Fortunately, providing proper nutrition for your flock doesn't require too much thought! Suppliers have formulated special feed complete with everything your chickens need. It comes in different formulations, such as starter for chicks, grower or developer for young birds, and layer for hens who are producing eggs. It also comes in different forms, such as crumbles or mash (referring to how finely milled it is). For adult birds, feed also comes in pelletized form. Some chickens prefer pellets while others prefer mash. Some don't care one way or the other, so long as it's good to eat.

And just for the record: If you wonder about hormones in chicken feed, there's no need to worry. There are no hormones in any brand of commercial chicken feed in the United States, according to the National Institute of Food and Agriculture (NIFA)—formerly the USDA Cooperative Extension Service, a nationwide non-credit educational network. Hormones are not, and have never been, allowed in the feed of even factory-farm raised chickens. We think the "hormones" rumor makes the rounds every so often because there are some brands of poultry feed labeled as "hormone free" while others don't bother. NIFA explains that it's illegal to use hormones in poultry feed in the United States.

However, because some brands of feed specifically advertise themselves as "hormone free," that begs the question for the other brands that don't mention it: "Do you use hormones in your feed?" But, rest assured, they do not.

Supplements

In addition to feed, you'll want to provide two supplements for your flock to consume as needed: The first one is oyster shell and the second is grit. They are *not* the same thing!

Oyster shell is a calcium supplement for layers. Although layer feed has a great deal of calcium, occasionally a hen will need more. If she has a lot of pasture, for instance, and is eating reduced amounts of commercial feed, she might need extra calcium. Just put it out in your coop, and let the chickens eat it as needed.

Grit is needed for proper digestion. Chickens don't have teeth, after all! The grit is stored in your chickens' gizzards, where it will help grind up seeds and bugs and bits of grass and greens. Chickens will need grit once they begin eating treats and food other than chick starter. If your chickens free range, they will probably pick up grit, gravel, or small stones on their own, but it's a good idea to provide extra just in case.

If your chickens don't get enough grit, their digestive systems can become clogged and impacted. Even if they do get enough grit, it can sometimes happen if you're feeding the wrong things. For instance, grass is great when chickens browse it on their own, but keep in mind that a foraging chicken will generally nip off only small pieces of grass, not long strands that can get wound up into a ball in her crop. So don't, for instance, feed your chickens long strands of grass left over from lawn mowing; the grass can tangle up in a knot in their systems and other food won't pass. An impaction means

When to Swap
Your Chickens' Feed

To know when to switch your chickens' feed, check the label of your particular brand of feed to see what the manufacturer suggests. Each manufacturer formulates their feed differently, so read the label and follow their instructions. Some recommend starter for only 4 weeks before moving onto grower; some combine both together in a starter/grower feed that can last up to 16 weeks.

That said, we have an additional caution: Don't give layer feed to chickens that are too young to lay eggs. The manufacturer's recommendations may suggest switching to layer feed at 16 weeks, or 22 weeks, or somewhere in between, depending on the brand. But keep in mind that those recommendations are general and aimed at breeds designed for early maturity and top egg production. If you keep breeds that are slow to mature or are primarily for show, 16 to 22 weeks may not be the best recommendation. Additionally, if your chickens were hatched in the summer and will come into maturity during the short winter days, it may not be the best advice, either.

Feeding your pullets layer feed when they are too young to lay won't make them lay any earlier, but it could cause skeletal problems from all the excess calcium. We recommend waiting until you see the first egg from your flock to change to layer feed. This may be a little later than the manufacturer recommends on the packaging, depending on the breeds you have and the time of year they mature. And certainly all your birds won't begin laying at once, especially if you keep many different breeds. But it is still a good general rule of thumb.

your chickens will need veterinary attention, or they could starve.

Free Choice

Some breeds may be able to barely subsist in good weather by free ranging (although this is unlikely, as chickens are domesticated animals), but most will naturally starve if you don't provide them with enough food. They will also be healthier and lay better if their bodies are not stressed out by undernourishment and nutritional deficiencies.

Our advice is to feed your birds at any age according to "free choice." That is, let them eat as much feed as they want and leave their feeder out at all times (although you may decide to take it up at night). Even if your chickens have access to pasture, free ranging simply supplements their diet. Chickens will eat as much food as they need to keep themselves healthy.

One of the worst mistakes a novice chicken keeper can make is to not provide enough food for their chickens, or to provide only scratch or corn rather than layer feed; scratch and corn just don't supply chickens with what they need to maintain proper health.

We do hear horror stories now and again from people who simply haven't done their research. One woman in Alaska called us to see if we could help her find out why her chickens were dying. It turns out she was throwing out a handful of scratch each day, and thought her chickens would be able to survive on what they could forage. Another customer wanted to know why her chicks were dying; we found out she was trying to force-feed them via a syringe, as if they were baby songbirds. And once or twice a year, we hear from people who've done research, but nonetheless choose to feed rolled oats, or cornmeal, or crumbled-up crackers to their flock, and then (unsurprisingly) have to deal with a host of nutritional and behavioral issues. Such issues include eggs with weak shells, egg eating by the birds, or no eggs—period. In addition, feathers can fall out; the birds can pick on one another or have loose, watery stools. And they can die in the cold or heat.

In the spring and summer your flock will probably eat significantly less feed than in the fall and winter, particularly if they have access to range. Still, they will need that base feed to keep them going. In the winter especially, they need calories to keep them warm, and in the fall, they need protein because they will be molting and renewing their plumage. In the spring and summer, they are usually at peak egg production, so all those calories go to help produce eggs and keep your hens in good condition.

In summary, **proper nutrition is the best thing you can do for your birds**; without it, you will have a host of problems. With it, you may wonder why everyone doesn't keep chickens as pets, since they're so easy!

Feeding Different-Age Birds Together

When you have older laying hens and young chickens together at the same time, keeping them eating the right feed for their age can be a challenge. With typical perversity, they are usually drawn to the feed you don't want them to eat. The older hens will want to eat the starter/grower feed that doesn't have enough calcium for them, and which also may be medicated. This medication is not only unnecessary for your hens, but it could be transmitted to you through their eggs. Meanwhile, the young chickens will want to eat the layer feed, which has too much calcium for them, and may cause potential skeletal problems.

But even when you have birds of different ages with different dietary needs, there is a way to manage. We recommend temporarily switching your whole flock to a grower or developer feed. Since grower/developer feed is unmedicated—check your brand, of course, to make sure—the eggs your girls lay while eating this can be eaten. This type of feed doesn't have all the calcium layers need, so it will not cause problems for growing birds who are not using it to produce eggshells. For your layers, you will want to make sure you are also offering free-choice oyster shell (separate from your feed) in case they need the supplementation for producing good, hard eggshells. Ideally, you should be offering oyster shell anyway, but it's especially important in this case.

Using grower/developer feed and offering free-choice oyster shell is a good compromise that will meet the needs of chickens who are at different levels of maturity and have different nutritional needs.

Mixing Your Own Feed

Did you know you can mix your own chicken feed? It's true! And you can find recipes online for how to keep it nutritionally balanced so your flock gets exactly what they need. But we don't recommend this. Usually you will want to stick to commercial feed, as it already has all the components measured and mixed in a perfect balance.

If you have found a good chicken feed recipe, it's not that it's a bad idea to mix your own, not in theory. It's just that it's not something most people can do well, and your birds' health and welfare depend in large part on the quality of their food. Not only is commercial feed sufficient—especially when your birds have access to pasture—it's also generally significantly less expensive than mixing your own (even when you're buying premium organic feed). Plus, if you're buying grain and other ingredients in bulk (including trace mineral mixtures, calcium, and kelp or alfalfa meal, for example), you often have to buy so much at a time that even if you have the capacity to store it all, it would get stale before a small backyard flock could ever go through that much.

If you're interested in mixing your own feed, by all means explore your options. You may live in an area where you can get all the ingredients easily. And it may become increasingly promising in the future to mix your own chicken feed at home, as keeping backyard chickens becomes more common and a support network grows. But whatever you do, be smart about it.

Mixing your own feed would have to be done very carefully so all the nutrients are balanced. Cracked corn plus barley and millet does not equal nutritionally balanced chicken feed; neither does sunflower seeds, flax, and alfalfa. Just any old mix of grains and adjuncts won't work. And despite what you might think, leftover game birdfeed or wild birdfeed does not equal nutritionally complete chicken feed, either.

Chickens have unique nutritional needs. Because they lay eggs so much more frequently than other domestic fowl, they need a lot of calcium, but too much can cause problems. They also need a lot of protein to lay their eggs and maintain their plumage, but too much protein can cause problems, as well. They need a specific mix of amino acids like methionine. And then there's the simple fact that they need a balance of vitamins and minerals just like other animals, including humans, so they don't develop deficiencies or even toxicities.

In 99 percent of cases, using a commercial chicken feed is the way to go.

COLD-WEATHER CHICKENS

If you live in an area with cold and snowy winters, encourage your flock to become good cold-weather chickens. Both you and your flock will be much happier for it.

Here are eight things *not* to do if your winters are cold and snowy.

1. **Don't keep your chickens closed up in their coop when it's cold.** Instead, good cold-weather chickens can be allowed to decide when they want to stay in or come out. You might think that your chickens won't want to go outside in the snow, and sometimes that's true. Some of your chickens will hate it and will stay inside most of the day, but others won't mind it at all, especially if you throw down hay or straw on top of the snow so it doesn't feel as cold to their feet. The only time we keep the coop door closed during the day is when the snow is too deep for our cold-weather chickens to walk in, or when it's just so bitter and windy we know no one will come out. (And, even then, we sometimes open the door just in case.)

2. **Don't tightly insulate your coop.** This seems strange, but it's true—tightly insulated coops can cause more harm than good. If your coop is tightly insulated, not only will it retain heat, it also will retain moisture—and retaining moisture in the coop is very bad. Chickens create a lot of moisture from breathing. A lot of moisture also evaporates from their droppings. And in winter, they'll be spending more time inside, even if it's just because of the longer winter nights. In a tightly insulated coop, all this moisture in the air can condense and freeze, contributing to frostbite. All that humidity also increases the risk of unhealthy conditions in the coop that can lead to respiratory ailments and mold-related illnesses. Plus, poor ventilation can also cause ammonia gas to build up inside your coop, which is damaging to your chickens' lungs. Cold-weather chickens need a coop to be well ventilated but not drafty.

3. **Don't heat your coop.** *Please.* This is another piece of advice that seems

completely counterintuitive. However, it's good advice for a number of reasons. Chickens adapt to lower temperatures over time. If the coop is heated, they'll never become real cold-weather chickens—their metabolisms will never acclimate to the cold winter temperatures outside. Then, if you lose power and their heat goes out, the sudden sharp drop in temperature means you could lose your whole flock in one terrible, fell swoop. Even if it doesn't come to that, if your chickens are hesitant to spend time outside, they will spend even more time inside the coop making the air wet and breathing the unhealthy, moist air. Finally, heating the coop is a fire hazard. Remember, chicken coops are generally pretty dusty places, and we hear stories every year from people who have lost their coops—and their flock (and occasionally even their adjoining garage and home)— to chicken coop fires. **The only time to heat your coop is during a sudden, precipitous drop in temperature,** just to help ease the transition for your cold-weather chickens. Complete losses due to heat cutting out and fires are among the worst catastrophes to befall a flock.

4. **Don't forget to gather eggs more often than usual.** If you have cold-weather chickens, some may continue to lay during the winter, and the eggs could freeze. While this doesn't really hurt them, exactly, it *is* a risk for bacterial contamination, because the frozen egg contents expand and can create tiny hairline cracks in the shell you might not see with the naked eye. The cracks can let bacteria into the shell. Of course, at cold temperatures, the bacteria don't grow very quickly, but nonetheless, keeping cracked eggs is just not a good idea. Plus, there's nothing like opening your refrigerator to find that a cracked egg has thawed and seeped out all over everything—what a mess!

5. **Don't let their water freeze.** Keeping fresh, unfrozen water for your flock in the winter can be a challenge. There are always heated waterers or water bases you can use, but even My Pet Chicken employees are split on their preferences. Some swear by heated waterers, while others just use multiple waterers. In the morning, you can carry a

Egg-cetera!

fresh eggs

Most hatchery Ameraucanas or Ameraucana/Araucanas are, in fact, Easter Eggers! Easter Eggers are a perennial favorite for their green eggs, hardiness, and friendly personalities. One clue that your hatchery may be selling Easter Eggers instead of purebreds is that they'll be labeled "not for show." Another clue—they may tout them as laying green or pink eggs.

fresh waterer to the coop and bring in the (now frozen) waterer that was in the coop overnight. By the time the one in the coop begins to freeze, your other one should be thawed; just switch them out again. Both methods work; one requires a lot of hauling and carrying, while the other requires the purchase of equipment. Determine which will work best for you.

6. **Don't put off coop cleaning.** Because your cold-weather chickens will be spending more time inside and creating more droppings inside as a result, the coop will need cleaning more often. You may prefer to use the deep-litter method discussed earlier in this chapter for managing your coop, rather than frequent cleanings, but even if you do this, you will need to add new bedding more frequently in the winter to make sure everything stays dry and sweet smelling.

7. **Don't let your birds get too bored.** If they have a very small coop and run, there may not be a whole lot to entertain your flock like there is during warmer months. When snow is on the ground, there will be little or no sunbathing. When the ground is frozen, dust bathing will be limited, too. There won't be bugs to catch or greens to forage. Bored birds may become snippy or even aggressive with one another if there isn't anything to think about or do other than reinforce the pecking order over and over again. Alleviate some of the boredom for your cold-weather chickens by adding treats to their area. For instance, hang

a head of cabbage in your coop for your girls to peck at. As they peck, it swings, making it more difficult to eat immediately, and keeping them entertained for hours. Suet cakes work well, too. In wintertime something that's also high in fat (like scratch or cracked corn) can help give them the extra calories they need to stay warm. You can also simply scatter some scratch inside the run for them to forage for—that will keep them entertained, too.

8. **Don't forget to protect their combs.** Most cold-weather chickens have small combs, but if you have breeds with very large combs, in the coldest weather they'll need extra protection. Spread a little petroleum jelly on their combs. Keep in mind that it needs to be a barely-there, thin layer. Healthy skin decreases susceptibility to frostbite, so you're just trying to keep the exposed skin from getting chapped and cracked. Some people assume if a little petroleum jelly is good, then more is better, but that's not true! Put on only as much as you'd want on your lips to help keep them moist.

HOT-WEATHER CHICKENS

Do you live in the hot, humid South or the scorching desert areas of the Southwest? If temperatures routinely reach 95°F or more in your area—and stay there for significant periods of time—you can help your flock withstand the heat by following these simple suggestions.

It's a lot easier to help your chickens out in cold weather than in hot weather. Chickens just generally deal with cold weather better. Plus, heat-hardy breeds aren't quite as popular with pet chicken keepers, because overall they tend to be on the flighty or active side.

Even so, here are six simple steps you can take to help keep your birds comfortable in excessively hot weather.

1. **Make sure they have access to fresh water at all times.** Chickens need a lot of water for an animal of their size—think about how much moisture goes into the production of an egg. In hot weather, they also need water to help keep themselves cool. Being without water during the day, even for a relatively short period of time, can throw their laying off and stress out their bodies, making them more vulnerable to illnesses, or even forcing them to molt.

2. **Keep the water cool, if possible.** Place frozen bottles of water in the reservoir of your waterer—or freeze water in muffin or bread pans to use. (The bigger the chunk of ice, the longer it will take to melt.)

3. **Make sure they have plenty of outside shade.** If you don't have trees or shrubs or shade from buildings or overhangs, you might rig a tarp over their run, set up a portable tent or gazebo, or invest in some shade cloth. Birds with dark or black feathers will especially appreciate a place to escape the beating sun.

4. **Assure that there is plenty of ventilation inside the coop.** On hot sunny days, the temperature inside the coop can skyrocket, and that heat needs a way to vent back out. Open windows or doors or use coop vents (making sure the setup allows your chickens to remain secure from predators).

5. **Give them hot-weather treats.** Your chickens will enjoy refrigerated treats like watermelon or frozen grapes and berries. In general, any sort of refrigerated scraps will be welcome. For a few additional recipes to share with your chickens in hot weather, see Chapter 15.

6. **Consider painting their coop white,** and using a white roof if possible, too, to reflect the heat.

Egg-cetera!

Chickens sometimes lay unusual eggs, ranging from eggs as tiny as a songbird would lay to huge eggs with more than one yolk. They may also lay eggs with pimples of extra calcium on the shell or even shell-less eggs. Young pullets are especially prone to laying irregular eggs.

GROOMING CONCERNS FOR NEW CHICKEN OWNERS

You must shampoo, condition, brush, and blow-dry your chickens daily, and clean and trim their nails every week. Ha ha! Just kidding. What a pain in the neck that would be!

Actually, chickens are awesome at taking care of themselves. There are just a few instances in which you'll want to groom your chickens.

Bathing Your Chicken

You'll probably never need to bathe your chicken. They clean themselves! If you are showing your chicken, though, or entering your bird into a competition, you may want to take extra care with grooming and appearance. You can find reference guides detailing the sort of things you should be concerned about in these situations.

On the other hand, if your chickens do get something stuck in their feathers that they are not preening out, you may decide to wash it out yourself. You won't need a book-long reference guide for a simple cleaning. Just keep in mind that shampoos can be drying to their feathers and make them brittle. So, if the dirt doesn't come out with simple water, try a pet bird shampoo. These are generally kinder to their feathers than what you might use on your own hair.

Make sure your bird does not get chilled when she is wet. If it is cold out when you want to bathe her, she should be dry before you put her back in the coop or yard.

And lastly, a word of caution: Most chickens won't like to be bathed, so be careful. Your bird may scratch or peck, even if she is normally calm, friendly, and docile. Bathing a chicken will also make a wet mess when she begins flapping her wings to object (and escape)—so be prepared. Having an extra set of hands will probably help. You may even want several extra sets, if you have never bathed a chicken before—and if she has never suffered the indignity!

Trimming Nails

You won't usually need to trim nails. They'll wear down on their own as your birds range on the ground. Chickens kept in cages or coops with wire bottoms are subject to foot and nail problems, but sometimes even if your birds have solid ground and run of the yard, their nails may grow too long. It's not especially common, but if you notice that your chicken's nails are so long that her toes can't take their proper position, you'll need to trim them so she can walk without difficulty and doesn't run into long-term problems.

Nail problems seem to be most common with broody breeds like Silkies, who may spend long periods brooding in a nest rather than foraging in the yard. It's a good bet that your bird will not enjoy the trimming process, but you will have a much happier chicken after you're done. You can use toenail trimmers designed for dogs' nails, or you can use some old toenail trimmers of your own. Your clippers simply need to be large enough to fit around your chickens' nails, and they need to be sharp enough so that the cuts can be made quickly and easily with no accidental twisting of toes.

When you trim, it's important to avoid cutting the vein in the nail, just like you would if you were trimming a dog's toenails. If your bird has dark nails, you might need a bright flashlight to stay away from the vein. Using a flashlight makes it a two-person operation, but you can more easily see where to stop. Stay ¼ inch or so away from the vein. Trim off a very small amount of the nail at a time—don't make a big cut all at once, or you are more likely to cut too far. If you pay attention, you can see if you are getting close to a vein because the color of the inside of the nail will change. If the color changes, you are too close—stop! Be very careful and don't push it. The nails don't have to be short; they just have to allow for the natural position of the foot. Also be sure to have some styptic powder or cornstarch on hand just in case you accidentally cause bleeding—you can use it to help the blood clot.

If you don't think you can manage trimming very well on your own (it is nerve-wracking, and there is a potential for trimming too far and hurting your chicken), you could contact a veterinarian or even a dog groomer in your area who has more experience.

Lastly, if you are seeing excessive nail (or beak) growth frequently in an active bird with access to the natural ground, it can sometimes indicate liver issues, and you may want to have her checked by a veterinarian, just to be safe.

Clipping Wings

It's preferable to let your chickens have their full, natural feathers. However, in some cases, feather clipping does make sense—especially if flying over the fence means ending up in the territory of a neighbor's dog, flying into traffic, or getting into some other dangerous situation. Clipping is painless for chickens when done correctly.

The nice thing about wing clipping, if it should be necessary, is that it only has to be done about once a year, after the molt—and it's a simple procedure performed with scissors!

There are three important wing-clipping considerations. First, **do not clip growing feathers or the feathers of juveniles.** Juvenile birds molt a few different times during their first year as they come into their adult plumage, so their feathers are frequently growing—and only feathers that have fully grown in should be clipped. Feathers that are currently in the growth cycle have a blood supply, indicated by a pink feather shaft, and cutting them can cause

Egg-cetera!

Those big white lumps in your egg on either side of the yolk are called chalazae. They're there to help keep the yolk centered in the egg.

your bird to bleed, sometimes excessively. Second, **clip only the first ten primary feathers**. This is all that is needed to prevent flight. Third, **remove only half the length** of those primary feathers. Any more than that is unnecessary.

Some people recommend trimming the feathers of one wing only, because this puts the bird off-balance for flight.

Finally, if you are squeamish or nervous about wing clipping, consider making an appointment with your local avian veterinarian. It should be a simple, inexpensive procedure, and you can watch and learn from a pro.

Wing clipping is simple if you have a helping hand.

Crest Trimming

Some chicken breeds like Silkies, Houdans, and Polish have large crests on their heads, and sometimes the crests can significantly obscure the vision of the birds. You don't *have* to trim the crest. But if you don't exhibit your birds, it won't hurt anything to trim the feathers, and it can help your bird avoid predators. It may even improve her station in the pecking order. After all, she can't avoid predators—or a peck from a dominant hen—that she can't see coming.

As with wing clipping, crest trimming will need to be done once a year or so. Be sure not to cut growing feathers. You will probably need a partner, too. One person can hold the chicken still while the other wields the scissors. We recommend safety scissors for your bird's protection—you will be trimming near her eyes, and she probably won't hold still for the process. Keep in mind that you won't need to trim the feathers close to her head; just give

her enough of a "hair" cut to improve her field of vision.

Silkies frequently benefit from a trimmed crest. While they don't enjoy the process, they are often very happy with the restored sight.

Beak Trimming

Ugh. "Beak trimming" is when a baby chick's beak is burnt or seared off in commercial farming operations to help manage the stress and aggression caused by close confinement in large factory farms.

Unless your bird is having some physical issue with her beak, beak trimming is unnecessary. Instead, provide your chickens with plenty of room so they don't pick on each other (refer to Chapter 5 for space recommendations).

There are some cases where trimming the beak can help your bird, though. For instance, if your bird is suffering from a condition called cross beak or scissor beak, trimming it may help her. This defect can be caused by the chick positioning herself incorrectly for hatching—normally, one wing will shelter the head inside the shell, but if she doesn't have her wing over her head, the skull can malform, and it will reveal itself within the first few weeks as a cross beak or scissor beak as she begins to grow. The tips of the upper and lower beak won't meet in the regular way; the beak will look twisted.

Even with this defect, in most cases, no beak trimming will be necessary. You will simply have to make sure the feeder and waterer are deep enough to accommodate the way your cross-beaked chicken must eat and drink, since it can be more difficult for her. Sometimes mixing a little yogurt into the feed will help her pick up the food. Other high-protein food such as scrambled or chopped hard-cooked eggs can help, too.

If her beak is growing unchecked, or if she seems to be having trouble eating, you'll need to consider trimming her beak. A veterinarian can advise you, and may also be able to show you how to do this, if he feels it's necessary. Again, this is different from the capricious, indiscriminate "trimming" done in factory farm operations to healthy birds, where practically the entire beak is removed. This kind of trimming simply addresses the tip (like trimming the tip of a

TRACI'S SQUAWKINGS

My favorite chicken of all time was a cross-beaked Easter Egger my husband and I named Buster. Her scissor beak wasn't apparent until she was 3 weeks old, but we weren't terribly concerned as she was growing happily along with all the other chicks. After a few months, however, her beak grew long enough and crooked enough that she just wasn't able to capture food. A visit to the local avian vet taught us that with a monthly beak trim and some hand-feeding, she'd be just fine. (And boy, was she a hit at the vet's office! Buster was their first-ever chicken patient, and no one there had ever seen a bird so tame and friendly. They even wrote her up in their monthly newsletter.) Until she was about 7 months old, I hand-fed her a fresh slurry of feed and yogurt two to three times a day, per the vet's advice. While it was time-consuming, I didn't mind. On the contrary, I felt privileged to be charged with caring for this vulnerable, sweet creature, and it was during the time spent hand-feeding that we really bonded. I loved her so, and I felt she loved me, too.

Over time, I realized that with a deep-enough food container, she could actually "bury" her beak and, using her tongue, draw pellets into her mouth. We experimented with a few different containers until we found one that was deep enough for Buster while also being heavy enough not to tip easily. It was a relief, on the one hand, not to have to hand-feed her, but on the other hand, I missed our bonding time. Buster ended up living to a ripe old age, happy and healthy until the end.

fingernail, but not to the "quick" or vein) to give your chicken a better quality of life.

The only *other* time you will want to consider beak trimming is in the case of an injury. Your chicken's beak can actually break or chip, although it is rare. *As long as it is just the tip*, her beak should grow back with no problems, and will probably need no attention from you. A chicken's beak will continue to grow throughout her lifetime. Normally, her beak will slowly wear down with use, but if her beak grows faster than it wears, the tip may break. A break can also be caused by an injury. Possibly your chicken caught her beak somewhere and broke it trying to escape. If that happens, check your coop and feeders to make sure there is no place to catch a toe or a beak.

A break at the very tip is probably not painful to her, although she may find it disturbing, since she can't pick up things in her beak the same way as usual. The break itself, so long as it is at *the very end* and she is able to eat, shouldn't present a long-term problem. Just watch to be sure she is eating and drinking as usual, since chickens will try to hide signs of pain and weakness.

If the break goes back far, it can be very painful indeed. If her beak was caught somewhere, she may have other bruising, too, from trying to get unstuck. A break like this may not grow back the same way, so it would be wise to consult a vet for large or dramatic breaks. If the edge is rough, it might be helpful for the vet to smooth it, in order to make eating easier and to help ensure her beak doesn't get caught elsewhere.

Keep in mind that sometimes liver problems, certain types of mites, or nutritional deficiencies can contribute to a weak beak. Your vet will be able to check out your hen and determine if there is some underlying issue. (The vet might be able to prescribe some pain medication, if necessary, too.) Even with a small break, contact a vet if you suspect there is a larger issue.

Dealing with Spurs

Only roosters have spurs, though every chicken, male or female, has little spur "buds." (Hens in their golden years can sometimes grow mini spurs.)

Roosters' spurs can be a real concern: Just because a chicken usually weighs 10 pounds or less—often *much* less—doesn't mean their spurs can't do serious damage. If you want to remove your bird's spurs, get the advice of a vet. It is too easy to injure your rooster, even break bones, if you don't know what you are doing. You wouldn't declaw your cat at home, would you?

If you must do something about spurs and don't have a veterinarian who can help, rather than removing them, you might decide to clip off the sharp tips and file them down a bit until they're blunt. A Dremel tool (used for woodworking) can make this a faster process. But you'll probably need a partner who can help you control your rooster while you're attempting to disarm him. The outside of the spur is like a fingernail, with no nerve endings, but the inside can bleed and hurt, like the quick of your nail bed, so be careful not to clip or file down too far. Your goal should be just to blunt the spurs. If you do that, your rooster will still be armed, but his weapons won't be quite so dangerous.

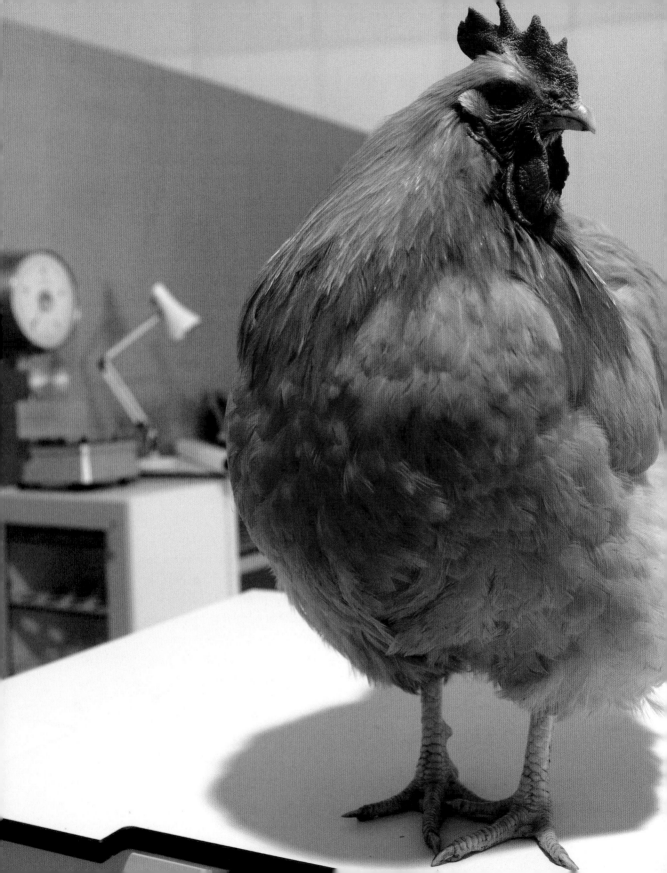

Determining When There Is a Problem

Even though you give your chickens the best of care, sometimes things can go wrong. In this section of the book, we'll offer some guidance on how to tell when something is amiss in your flock and how to handle illness and other issues.

Just like new parents, new chickenistas often excessively fuss and fret over their flocks. Which chicken behaviors are normal? What signs and symptoms indicate that something has gone wrong? In this section, you'll learn how to discern between the two and what to do if one of your birds *is* having a problem. We'll go over biosecurity basics, and we'll teach you how to "think like a chicken" to make it a breeze to add newcomers to your flock.

RECOGNIZING SIGNS of ILLNESS

As pets, chickens are easy. They're friendly, cheerful,
and productive, and they don't take a whole lot of time.
But just like any other pet, they can get sick
and may require veterinary care.

Chickens don't need daily walks on a leash or regular grooming. They don't leave fur all over the rug. They don't sharpen claws on furniture or chew up your shoes. But the challenge with any pet, not just chickens, is determining when they are feeling ill, since they can't tell you in words. Chickens in particular actively try to hide signs of illness. How frustrating!

Why would they hide signs of illness? Chickens are prey animals, meaning they're on the lower end of the food chain. Most predators will enjoy a chicken dinner—and your chickens know this instinctively. For that reason, they'll hide any weaknesses that would attract predators looking for the easiest catch.

A sick bird can experience negative conse-quences even from other members of her flock. If the other girls realize Matilda is too weak to effectively defend her place in the pecking order, the hens below her may make a move to depose her. Plus, the flock will occasionally try to drive away sick members. Perhaps this is a mechanism to keep contagious disease from taking hold; perhaps it's an issue of not wanting to attract opportunistic predators to an easy lunch. But whatever the case, a sick chicken will not usually broadcast her illness. Instead, she will often stay inside the coop or in another hidden spot where you won't be able to observe the changes in behavior that could clue you in to the fact that she's under the weather.

Even so, most backyard chicken owners are pretty tuned in to their pets' personalities and can tell when a bird may need attention, even if they can't make a firm diagnosis like an avian veterinarian can. For instance, general signs of illness in chickens include lethargy, inactivity, standing or sitting with feathers ruffled, droopy wings, difficulty breathing, a pale comb, and changes in droppings. Indicators like these suggest that your hen may not be feeling well, but they don't necessarily indicate what illness your chicken has. An examination by a vet will be needed to determine what could be going on.

Keep in mind that your chickens may exhibit signs that aren't related to an *actual illness*. Birds may have watery droppings when it's very hot outside, because they are drinking extra water to help keep themselves cool. Or they may be eating something that changes the color of their droppings, like berries. Your hens may seem lethargic and sit around with their feathers ruffled—but they are broody and wanting to hatch eggs, not ill!

So, how can you determine if your bird is ill or not? First check out our handy chart.

Common Signs of Illness in Chickens

SIGN	POSSIBLE HEALTH PROBLEM	POSSIBLE BENIGN CAUSE
Ruffled feathers	Many illnesses	Broody, hot-weather response
Panting	Various respiratory illnesses, gapeworm	Hot-weather response
Drooping wings	Coccidiosis and other illnesses	Hot-weather response
Lethargy/inactivity	Many illnesses	Sleepy, relaxed, sunbathing, cold, hot, broody
Loss of appetite	Many illnesses	Eating secretly, foraging well, broody
Feather loss	Nutritional deficiencies, parasites	Broody, molting, pecking order issues, mating
Drop in laying, shell issues	Many illnesses and nutritional deficiencies	Broody, young, stressed, egg hiding or eating
Pale comb	Many illnesses and infestations	Broody, molting, seasonal break in laying
Spots on comb	Fowl pox	Normal scab from pecking or a scratch
Coughing, sneezing	Various respiratory illnesses, gapeworm	A physical object stuck in nares or throat
Changes in droppings	Coccidiosis and other illnesses	Hot-weather response (watery droppings), unusual diet
Lump or tumor	Crop impaction, various illnesses	If it's on the front-right part of the breast, it's your bird's crop
Tremors, neurological signs	Various illnesses, nutritional deficiencies	Occasionally benign, but see your vet
Nasal discharge	Various respiratory illnesses	Unlikely to be benign; see your vet
Swelling/bubbles in eyes	Various illnesses	Unlikely to be benign; see your vet

PERFORMING A HOME EXAM

If you think that your chicken might be ill, gather more health information by doing a home exam. Here's what you will need to do.

• Check for wounds first; they can sometimes be easy to miss under all those feathers.

• Examine the color and state of your bird's comb and wattles. Are they red and full or pale and shrunken? Is there a rash? Are there a lot of black spots?

• Look carefully for infestations of lice or mites—they especially like to congregate under the wings and around the vent.

• Be certain that her leg scutes (scales) are smooth and clean (dirty, raised scutes can mean she is suffering from scaly leg mites).

• Does she feel heavy or is she losing weight?

• Are her feathers clean and glossy, or do they have a mangy appearance, especially around the vent?

• Evaluate the fullness of her crop. If it's full, is she eating? Or is it too full and unable to empty?

• Palpate to see if she has a hard mass in her abdomen that might indicate she is egg bound (unable to pass an egg).

• Check her vent to see if it is in good condition.

• Determine if she is breathing easily, or if her tail or body is rocking up and down with the effort.

• Make sure her eyes and nares (nostrils) are clear.

• If you know which eggs are hers, do they seem normal? Are there shell issues? Is she laying at her normal rate? Are her eggs their normal color?

• Last, examine her droppings—do you see any worms? Are the droppings normal in color and consistency?

Once you have made your examination, review what you have learned about the hen's condition. Some things are relatively easy to take care of at home, like mites, lice, worms, and even most wounds. Other conditions, such as a suspected impaction in her crop or reproductive tract, will require a visit to the vet.

Signs of Special Concern

You should be especially concerned about serious illness if you observed any of these signs.

• Consistent coughing, wheezing, sneezing, or nasal discharge

• Swelling around the eyes, neck, or head, or bubbles of goo in the eyes or nares

• Long-lasting purple or blue discoloration of the wattles, combs, or legs (disregarding it, of course, if it's a breed that sports blue legs, a mulberry comb, or other unusual coloring)

• Lethargy and lack of appetite paired with watery, green diarrhea

• Drooping wings paired with tremors, circling, twisting of the head and neck, or lack of movement

If any of your birds are exhibiting these last signs or if you are experiencing an unusual number of unexplainable deaths in your flock, call your local veterinarian, your state's veterinarian, your state's animal/poultry diagnostic laboratory, your local cooperative extension office, or the USDA Veterinary Services office for more information and advice.

The key is to remember that you shouldn't panic the first time you notice anything unusual. Use common sense: If what you're seeing could be benign and normal—like broodiness—then take a few hours and make additional observations before you make an emergency appointment with the vet. And if you've determined your bird is broody? Well, still keep an eye on her. After all, it is possible for a hen to be broody *and* sick. For instance, broody hens can be an easy target for mite infestations, since they don't normally dust bathe while brooding and they spend all day, unmoving, in the nest.

If, after having spent some time gathering information, you determine a call to the vet is necessary, you'll have that much more to share when the appointment comes. That additional information will aid with diagnosis.

Other Helpful Information for Your Vet

Your vet will want more information than just symptoms to determine what could be wrong. You'll also want to be ready with basic background information about your flock.

Some things your vet may need to know:

• How old is your flock?

• If they are chicks, how warm are you keeping the brooder? Is there a place for them to escape the heat if necessary?

To prevent the spread of illness, immediately quarantine any birds you suspect may be unwell.

• Are there any that are not growing as quickly as the others?

• If they are adult hens, how long have they been laying?

• Has this hen had a drop in production recently, or any problems with shells (thin, wrinkled, pimpled)?

• Has her place in the pecking order made a sudden, drastic drop?

• Has there been any excessive fighting in your flock recently?

• Has there been a drastic change in weather?

• Does their outdoor area stay dry, or does it get muddy?

• Have there been any unexplained deaths in your flock recently?

5 Signs Your
Flock Is Stressed

Your flock can become stressed out and unhappy, and the signs aren't always obvious. They're not chewing on your furniture to try to relieve separation anxiety like your dog is, and they're not peeing in the laundry room like your cat is. Here are some signs your chickens may be stressed.

1. **Illness.** Sick hens are unhappy and stressed. Duh. But the reverse is true also: If they're stressed and unhappy, they will be more vulnerable to illness. This means that if you're seeing a lot of sickness in your flock—or a lot of problems with egg laying, such as thin shells or oddly shaped eggs—there could be something in their environment that is either taxing their immune systems or just plain stressing them out. See if you can determine what it is. Check the security of your coop and run, as well as the amount of space you're offering. Make sure there is enough high-quality feed for everyone. Is your dog barking at them? Even if they're safe behind secure wire, the harassment could be taking its toll, and the stress of it all could be leading to illness or egg problems.

2. **Egg eating.** Eggs are yummy—we don't deny it! They're a great source of vitamins and protein, too. Even for chickens, as disturbing as that thought might be. However, when hens are eating their own eggs, it's generally a sign that they're suffering from a nutritional deficiency, often of calcium or protein. Or it could be that they're bored and stressed, without enough space. Possibly eggs aren't being gathered frequently enough—or, perhaps, there aren't enough nesting spots for everyone. If you can't determine what's causing the problem, consider having a veterinarian or experienced poultry-keeper come to check for anything amiss.

3. **Aggression.** Unhappy hens will often pick on one another. Humans do that, too. If you're grouchy or you've had a bad day, you may take it out on an innocent person. Hens will take it out on anyone lower in the pecking order than they are. The girl at the low end may get especially ragged in times of stress. If you're seeing a lot of pecking and aggression in your flock, again, just check to see if there is some sort of stressor you can remove that is causing the bad behavior.

4. **Listlessness.** A happy hen will spend her day foraging and scratching in the yard, sunbathing, dust bathing, finding a place to lay her egg, and generally just hanging out with her flockmates. If your hen isn't doing these things, something is wrong. It could be that you just have a broody hen—broodies will want to stay quietly in the nest. If you've determined that it's not nesting behavior, though, you need to see if there's something going on. If you conclude that your hen is ill, consider quarantining her right away.

5. **Crying.** Hens that are very unhappy may cry and whine, just like your dog or cat might cry when it's unhappy. It doesn't *necessarily* mean anything is seriously wrong. It may be heartbreaking to hear, but sometimes it's just a part of owning pets. Just because your dog is whining for a piece of cheese, that doesn't mean there's something in his environment you need to correct. If the whining and crying aren't motivated by something you understand, you should be concerned and try to find out what's going on.

- Have you had any predators in your area recently—could your bird be wounded?

- Have there been any other illnesses?

- What have they been eating?

- How fresh is their food? Has it gotten wet? Could it be moldy?

- Are you offering grit and oyster shell?

- Are you offering any treats? If so, what and how much?

- Is their water fresh? Has it run dry or frozen recently?

- What type of bedding do you use?

- When was it last changed?

- Could your flock be eating something unusual?

- Do they have access to the ground and to pasture?

- Have they been exposed to wild birds?

- Have they been to poultry shows recently where they would have been exposed to other chickens or game birds?

- Have you had any visitors to your property who also keep chickens? If so, what biosecurity measures did you use during the visit? (See page 165 for more on biosecurity.)

- Have you added any new birds to your flock recently?

- If so, were the new additions from an NPIP flock? What quarantine did you provide, if any?

All these things will help your vet determine what the problem might be. Without this detailed information, uncovering the problem will be dif-ficult at best. In most cases, the vet will have to examine the bird to make a diagnosis.

REASSURANCE FOR NEW CHICKEN KEEPERS

Here are 16 "symptoms" that commonly worry new chicken keepers—but shouldn't.

1. **Small eggs.** In a new layer, small eggs are the norm. It takes a while for your chicken's eggs to reach the full-potential size for their breed. If they are yolkless, those tiny eggs are known as fairy eggs.

2. **Eggs with thin or no shells.** Your chickens' first eggs will be pretty pathetic! Not only will they be small, it's also common for shells to be weak at first. Some eggs may not even have shells. If it's not a long-lasting problem, it's no reason to worry. This is not a sign of sickness; it just takes a while for your hen's ovulation cycle to stabilize. In fact, you should be celebrating! Your young pullet has just come of age.

3. **Occasional egg issues in an older layer.** Some eggs might have wrinkled, rough, or thin shells. Occasionally there will be a calcium "pimple" on the outside of an egg. Some eggs may be smaller or larger than normal. Changes in eggs can happen when there are disturbances in the coop, such as a loud thunderstorm, or if your flock has been harassed by a predator, like a dog or raccoon. They can sometimes happen due to

the stress of new birds being introduced to the flock, or to other chicken-y stresses we may not notice, even with close observation. So long as egg issues are occasional, don't be worried.

4. **Annual feather loss in late summer or autumn, paired with a drop or cessation in laying.** Molting occurs once a year in mature chickens; they lose their feathers and regrow new ones. It's their way of refreshing their plumage, just in time for the winter so they'll have a "new coat" for the season. They can go through either a hard or a soft molt. A hard molt means all the feathers are lost almost at once, whereas with a soft molt, the feathers are lost and regrown gradually, so it may not even be noticed. Sometimes molting chickens look bald or mostly bald, depending on the stage. Sometimes they look like porcupines. How long it takes for feathers to regrow depends on the bird, the feed, and other factors, but just to give an idea, 6 weeks would not be surprising.

5. **A little blood on the outside of the egg.** Such spotting occasionally happens if a capillary breaks in the reproductive tract while your hen is laying her egg. If it becomes frequent, or if there is a significant amount of blood, check with your vet.

6. **A blood spot inside the egg.** A blood spot isn't the beginning of a chick, and you don't have to have a rooster in your flock to have a blood spot. This kind of spot can happen if a capillary breaks a little farther up the hen's reproductive tract, before the shell has been laid over the albumen-enclosed yolk. It's unsightly, but nothing to be worried about. It can be consumed, or you can remove the spot with a spoon if it disturbs you. (Supermarket eggs are candled before sale to remove eggs with blood spots.)

7. **A lightening of eggshell color.** The color of your chickens' eggs will lighten up over the course of the year. It happens more dramatically in some breeds than in others—for instance, Marans' eggs will often get considerably lighter in color by the end of the year. Unless this change is accompanied by respiratory signs, it's nothing to worry about.

8. **A lump on your chicken's breast.** Many new chicken owners panic when they see that their new chicks have "suddenly developed tumors!" The suspected "tumors" are most likely just the chicken's crop—a part of the digestive system, which is located on the front-right part of the breast. It will be smaller in the morning before she has eaten, and larger after she is full. The crop holds food before it goes to the gizzard to be ground up with the grit she eats. If the lump doesn't change in size or is hard all the time (don't press too much, because you can force the food back up into your chicken's mouth and cause her to aspirate her food—very dangerous!), your chicken might have an impacted crop, and you will need to get her to a vet.

Two hens enjoy sunbathing on an early autumn morning.

9. **Convulsing chickens.** This sounds terrible, doesn't it? But before you lose your mind with worry, look to see if your bird is just dust bathing. Chicks will also "dust" bathe in the shavings of their brooder. It is not a graceful process, and more than one chicken owner's heart has skipped a beat or two the first time they saw dust bathing.

10. **Chickens laying down with one wing stretched out as if dead.** It's scary! But more than likely they're just sunbathing (or, as chicks, heat-lamp bathing). It's nothing to worry about.

11. **Refusing to leave the nest.** In most cases, this is broodiness. Your hens may go broody at any time. This is when they stubbornly insist on sitting on eggs in order to hatch them into baby chicks. It doesn't matter if the eggs are fertilized or not; some hens will even go broody on golf balls, old doorknobs, or wooden eggs!

12. **A black spot on the comb.** When a chicken gets a peck or scrape on the comb or face, this will manifest as a black spot or scab, and isn't usually a cause for concern. If *all* of your chickens seem to have rashlike spots, if they are yellow or show signs of pus, or if you are sure the spots you see are not from a scrape or peck, you should be concerned. But a single scrape or two? Don't worry.

13. **Watery whites.** The egg whites in eggs laid by older hens may be watery. This is normal. It is also normal to have watery whites in the eggs laid by your birds during a heat wave.

14. **Cloudy or discolored egg whites.** Cloudy whites in a farm egg are usually indicative of its freshness. The cloudiness is caused by dissolved carbon dioxide in the white. You don't see this in older eggs because, over time, the carbon dioxide escapes through the shell. If the white of the egg is greenish, this usually means there is too much riboflavin in the diet. Make sure your birds are getting balanced nutrition. Eating acorns, shepherd's purse, or some other types of weeds can cause that greenish tinge, too. Pinkish whites are caused by some types of weeds a chicken may get into, or high quantities of cottonseed meal in the diet. Pink or green albumen may also be caused by pseudomonas, a bacterial infection. In this case, isolate the affected chicken right away and get it to the vet.

15. **Fishy-smelling eggs.** Don't worry about illness in this case—but do worry about diet. In some hens that lay brown eggs, eating too much canola or rapeseed meal can cause a fishy smell in the eggs. Not all hens are affected by the process that causes the smell. The smell is caused by the accumulation of trimethylamine (TMA) in the yolk. Most hens metabolize the TMA into another (odorless) compound, but brown egg layers don't do that as efficiently, so when feeding canola or rapeseed meal, in some cases, you may end up with—ick—fishy-smelling eggs.

16. **Sticky poo that looks like melted chocolate.** Chickens actually produce two types of poo: fecal poo and cecal poo. Cecal poo is thicker, stinkier, and stickier. It usually looks sort of like melted chocolate, and it occurs once every eight or so poos. It is normal and nothing to worry about. Some foods may cause sticky poo, too. For instance, lots of barley in the diet can cause sticky, tarlike poo. Barley contains substances that chickens find difficult to fully digest, and it can cause very sticky, viscous-looking poo. Sometimes various fruits will cause looser, darker poo from the sugar and extra moisture. If your chicken's diet has been different as of late, you may want to revert to fewer treats to see if that is the problem.

Remember, the most important part of keeping your chickens healthy is providing a spacious, hygienic environment, so be sure to keep

Egg-cetera!

fresh eggs

All chickens have spur buds, but (with a few exceptions) only a rooster's spurs will grow to any length. At what age this happens varies widely from breed to breed. Some can start growing as soon as 3 months, while others take 7 to 8 months to develop.

the coop clean and give your birds the recommended amount of space. Overcrowding can cause stress, which will make them more susceptible to illnesses. Since you'll be checking on your birds daily—sometimes several times a day—ideally you'll be able to catch any problems early, which will increase the chance of a positive outcome.

BIOSECURITY PRECAUTIONS TO TAKE WITH YOUR FLOCK

Most illnesses or even infestations (such as mites or worms) are contracted when your birds are exposed to other birds, either directly or indirectly. For that reason, there are simple steps you can take to reduce the chances of exposure and keep your flock healthy.

1. **Use common sense and restrict access to your birds and to their area.** It's fun to have chicken-keeping friends, but remember that allowing visitors to your flock who have been exposed to other birds is one of the main ways illnesses and infestations can get passed from one flock to another.

2. **Before and after working with your birds, simply wash your hands.** Use regular soap and water—antibacterial soaps are no more effective than regular soaps—followed by a disinfectant, like an alcohol-based hand sanitizer.

3. **Clean, then disinfect, all new equipment that will come into contact with your chickens** (such as feeders, waterers, nests, toys, roosts, shovels, brooder enclosures, etc.), and when the equipment gets soiled from use, clean it again.

4. **Dedicate a special pair of shoes for use when working with your chickens, or clean your shoes with disinfectant before and after working with your flock.** Why before? Remember, you can carry things into the coop on your shoes, and you don't want to bring any illnesses home from the feed store, the trip to the park to feed the ducks, your friend's flock, or anywhere else.

5. **Don't share tools or equipment with other poultry owners,** unless you are willing to thoroughly clean and disinfect those tools with every use.

6. **Make sure you buy your birds from a reputable hatchery or dealer, preferably one that is NPIP certified.**

7. **Always quarantine any new birds you acquire from the rest of the flock (see the following section for more on this).** While this isn't an issue with day-old baby chicks purchased from major hatcheries here in the United States, it is important when acquiring older birds, particularly rescues.

Why You Must Quarantine New Birds

One of the biggest challenges you'll face as a pet chicken owner is introducing new birds to your established flock. Not only is it an issue of behavior (see "How to Introduce Newcomers to Your Flock" on page 177), but it's also a matter of biosecurity. So, when introducing new birds—*especially* if you've purchased adult birds or adopted a stray—it is critical that you follow biosecurity recommendations and *quarantine them for at least 4 weeks*. This means you'll want to keep the new birds away from your flock entirely. You'll need this time to ascertain if the new birds have any infections or communicable health issues that the rest of your flock can catch.

Clearly, an unscrupulous person might sell you a sick animal without telling you—and that's a problem! Quarantine might help you spot the problem before it affects the rest of your flock. But quarantine is not suggested solely to protect you from the machinations of scoundrels, so don't assume that any sick chicken was just a purposeful attempt to swindle you. With chickens and their propensity to hide illness, it may well be that the person who sold you the sick hen had no idea she was sick. Or it could be that the illness wasn't yet at the point where signs were observable, even for the most experienced chicken keeper. Or you could just be dealing with a beginner or someone who is not adept at spotting potential problems. That's why, even if you're getting a bird from your best friend or your mom, you should quarantine. It isn't a matter of trust; it's about taking mature, appropriate precautions.

This goes double with rescues. When you see a hen or rooster in need, your first caring instinct may be to give that chicken a new forever home. But your rescued bird may be hiding a contagious illness or infestation. You don't know where the bird has come from or what she's been exposed to. We're not saying you shouldn't rescue a chicken in need! What we are saying is that you should be wise enough to use quarantine methods to protect the rest of your flock. You quarantine because, if the new bird is ill, you can get a diagnosis and treat it while protecting your established flock.

If the bird is sick, it might be something relatively easy to deal with at home, like mites or lice. But it could also be something more serious, something that requires intense veterinary care and medication, injections, or antibiotics. Or worse, it could be something fatal. How terrible would it be for an act of genuine compassion—trying to help a bird in need—to turn into

Egg-cetera!

fresh eggs

A blood spot in an egg doesn't mean the egg is fertilized. It just means a capillary broke somewhere inside your hen's reproductive tract while the egg was being produced. It's unsightly, but fine to eat. Commercially produced eggs are candled to remove eggs with blood spots.

When ill birds are picked on by the rest of the flock, they may try to hide themselves.

something that destroys your own flock, or causes them to suffer?

That's why you must quarantine newcomers. This means they should not be in the same space together—not the same run, not the same coop. They shouldn't be breathing the same air. You may decide to keep your new additions inside your house for a while, or perhaps you have a small hospital coop that can keep your new birds safely separated from the rest of your flock for the quarantine period.

During this time, you'll want to observe your newcomers carefully. Look for signs of infestation: creepy crawlies like mites or lice, or raised leg scales, which might indicate a scaly leg mite infestation. Look for unusual droppings. Are they foamy? Bloody? Yellow? Look for unusual behaviors—does she sit with her feathers ruffled all day? Is she lethargic? Is there any coughing or sneezing? Is there any discharge from the eyes or nares? Are there any skin conditions?

If the new birds are sick or carrying parasites, treat them and make sure they are fully recovered before introducing them to the rest of your flock. If they have a communicable disease, be sure to consult with your veterinarian to find out when it will be safe to begin introductions, because some illnesses can be communicated even after recovery.

When introducing day-old baby chicks from a major NPIP hatchery, you probably won't have to worry about strict quarantine. Most states require baby chicks and juvenile birds that ship across state lines to originate from hatcheries like ours, hatcheries that participate in the NPIP, meaning they incubate their eggs in facilities that are biosecure, and are often many miles removed from the breeding flocks. This means that day-old baby chicks from NPIP hatcheries are not a risk for your established flock.

OTHER WORRIES

The first year you keep chickens, there may be a steep learning curve. You want to do what's

Dealing with
Mites and Lice

Mites and lice are a common problem for chickens and can be carried in by wild birds, so if you have an infestation, it doesn't necessarily mean your management practices are bad. Infestations can cause a reduction in laying, pale combs and wattles, anemia, and even death. Infestations can also cause feather loss, usually on the back, because a bird may excessively preen and pluck her own feathers in an attempt to get relief. It is similar to the loss of feathers caused by too much attention from a rooster. However, a rooster normally causes broken feathers, at least at first, whereas mites can quickly cause feather loss down to the skin; sometimes the skin is irritated and red. If you have mites or lice, you will need to treat all of your birds.

Mites inhabit your birds' feathers and suck their blood—yuck! Luckily, the types of lice and most mites that affect birds are not interested in humans. A mite is so tiny that it's hard to see, and some types of mites only come out at night. They are usually dark- or red-colored after they have eaten. If you have good vision, you may be able to see them around your bird's vent or beneath her wings, especially if your bird has light-colored feathers, which make it easier to see the mites against. Lice are larger, but they're lighter than mites, so they're nearly as hard to spot.

There are various treatments for these avian parasites, but they more or less all involve using a pesticide dust or spray on your birds. This sounds sort of terrifying, but think of it like a flea dip for birds. For a non-chemical solution, food-grade diatomaceous earth can help prevent and treat some types of mites and lice; however, it works slowly. We do recommend using a few handfuls of diatomaceous earth in your birds' dusting areas prophylactically to help prevent infestations.

If the infestation is severe, you may want to use something that will kill pests on contact to provide your birds relief. You can use either a dust or a spray. We prefer the dust treatment because the girls enjoy the feeling of dust in their feathers, whereas they dislike getting wet. Be careful when applying the dust, though, that neither you nor your birds breathe it in. We have used old spice shakers (clearly marked POISON). Other people prefer to fill the toe of an old pair of panty hose with the powder and pat it on that way: The dust is so fine it filters through like a powder puff. When you apply the treatment, pay particular attention to the areas around their vents and beneath their wings. If you have a rooster, make absolutely sure to treat him thoroughly, because if he has mites, his amorous attentions will spread it back around to the rest of your girls.

Hold your birds firmly as you apply the powder: They may react to your kind ministrations by flapping their wings vigorously, making a real dust cloud that you don't want to breathe in.

best for your flock, so you will be highly attuned to anything out of the ordinary—but with no baseline of "ordinary" established, you may not know when you should be worried. Here are a few things that should raise concern.

Leg Mites

If the scales or scutes on your chickens' legs are inflamed, crusty, or painful looking, your chickens might have scaly leg mites. The scales on their legs may begin to lift up due to waste from the mites building up under the scales. The condition is painful and eventually debilitating for your birds. Imagine having your fingernails slowly forced up, day by day. Birds with feathered legs tend to be more vulnerable to this kind of infestation, since the feathers on the legs can give the mites easier ingress.

The good news is that leg mites are among the easiest things to treat at home. You can treat your chickens for scaly leg mites by rubbing their legs with petroleum jelly or even an antibiotic ointment that uses petroleum jelly as a base. Some people prefer to use oil or even a nonstick cooking spray. You will have to treat this condition every 2 or 3 days for several weeks, and the scales may not return to normal, but you will be able to get rid of the mites.

If your chickens don't like to be handled, it's usually easiest to wait until dusk when they've settled into the coop for the night. They're easy to catch at this time, and if they've had a little time to settle into sleep, they probably won't resist handling at all.

Feather Loss

Is feather loss something to be worried about? Sometimes. Feather loss can be caused by a number of things, so to determine if you should be concerned, let's go through the list.

The first thing that comes to mind is molting. When feather loss happens all over the body, the bird may be molting. Molting is nothing to be worried about; your birds will molt annually.

If it's not molting, the next thing that comes to mind is stress: Happy chickens don't pick on one another, but stressed chickens do. Pecking, also called picking, is almost always the result of high stress levels. When it happens, the birds will sometimes pluck each other's feathers out, which can hurt them. Those lowest in the pecking order may have bare spots on their backs or heads.

To deal with this problem, you have to figure out why your chickens are unhappy. Do they have sufficient space? Try letting your chickens free range, and see if that solves the problem.

Are there enough feed and water containers to go around? If the birds don't all have enough access to feeders and waterers, give them additional feeder space so they don't have to compete for food. Also make sure you are letting them eat "free choice." Don't restrict the amount of food they get every day, or only those highest on the pecking order will get to eat.

Are there enough nesting boxes? If someone is broody and won't come out of the nest, there may not be enough space for your other girls to lay.

If you have roosters, are there enough hens to go around? Each rooster should have 10 or so hens, so the hens don't get overbred. Even when do you have enough hens, it's possible your rooster may have a favorite hen. If that's the case, you may want to get a hen saddle, which can help protect your hens' backs from overzealous or excessive mating. Feather loss

Worming Your Chickens

Chickens with intestinal worms require treatment. Signs that may indicate the presence of internal parasites include pale combs, a drop in laying, and watery droppings. Feather loss can also occur if the infestation is severe. (Parasites of any sort can drain a chicken's system and make her feathers brittle.) However, these same signs can indicate several different illnesses: Seeing them doesn't automatically guarantee a diagnosis of worms. As a responsible pet owner, you'll need to investigate further and consult a vet to find out if your chickens have worms and, if so, what type they have.

Obviously, you do not want to treat for worms if your chickens don't have them. First, if they are sick, an unneeded medication will be a strain on an already-taxed system, potentially making them sicker. Second, treating your flock for worms when they're suffering from something else will also delay the administration of treatment appropriate to their actual condition—again, making them sicker. Plus, if your birds do have worms, you want to make sure to use *correct* anthelmentic (wormer) for their specific infestation. For instance, tapeworms and roundworms are treated by different wormers, so if your chickens have roundworms, using a medication for tapeworms will not address the problem they have and could make the situation worse.

Your veterinarian will be able to perform a fecal smear and tell you which parasites your flock is suffering from. Avian vets can be hard to find, but any vet can perform a smear for you, whether they treat birds/chickens or not. Local cooperative extension agents will sometimes make this service available as well, so in most cases, you can find someone to do a smear for a nominal fee.

If the fecal smear is negative for worms, consult with your vet to help determine the cause of your flock's signs of illness. If the fecal smear is positive, then your vet will recommend a course of treatment based on the results—treatment that you can be sure will help, not hurt, your flock.

This is important: Be sure to ask if you need to discard eggs while your chickens are being medicated and, if so, for how long that should continue. The good news is that some new wormers, specifically Hygromycin B, treat several types of worms and don't have an egg-discard period!

If all your detective work doesn't help, consult with your vet for help in determining what's causing the feather loss.

caused by mating occurs mostly on the back, or the back and neck.

If the feather loss is not being caused by stress or roosters, could your flock be suffering from mites, lice, or other external parasites?

Pregnancy and Chickens

There's a rare condition in which chickens can actually become pregnant! Just kidding. We're talking about keeping chickens while *you're* pregnant.

Think there's a lot to worry about if you want to keep chickens while pregnant? Think again. There is no reason you can't keep chickens when you are pregnant—not if you use common sense. Farm wives have done it for time out of mind. Just use the same common sense sanitation precautions you would use when keeping other animals: Wash your hands after handling them, collecting eggs, or dealing with bedding and other chicken care activities. Avian and human metabolisms are quite different, so there are few things that affect both humans and birds. Even lice that are interested in your chickens are not interested in humans.

However, just like cats, chickens can get toxoplasmosis, which can cause a mild illness in human mothers exposed to it, but can cause serious harm to a fetus infected *in utero*. This is the reason pregnant women are advised not to clean a cat litter box while pregnant. So, for that reason, have your partner or hire someone else to clean the brooder or coop while you're pregnant, just to be on the safe side. Your doctor or midwife may have other recommendations based on your particular circumstances. If that's the case, be sure to follow her advice on the subject.

Should I Be Worried about Salmonella?

Worried? No. Informed? Yes.

Humans do not catch salmonella from chicks or chickens the way you would catch a cold from your neighbor. Salmonella is food poisoning; you get it from eating infected meat or eggs that have been improperly prepared, that is, not completely cooked through. Or you can also get salmonella by contaminating your hand or an object with feces and then putting it in your mouth. People more at risk for contracting salmonella are either very young, very old, pregnant, or have immune systems that are already compromised in some way. The best way to keep your children and family safe from infection is by keeping your own hens, whose conditions you can monitor, and by having your family members wash their hands after dealing with chickens. You want to have them wash their hands after dealing with any pets, for that matter. Alcohol, either in a sanitizer gel or from a bottle of rubbing alcohol, is an effective sanitizer for salmonella bacteria.

Nonetheless, every so often it will be in the news that some people have gotten ill with salmonella from eating vegetables, such as tomatoes or spinach. In these cases, the food was usually contaminated by rodents in a warehouse somewhere. You also may have heard about salmonella outbreaks at several factory egg farms.

When it comes to issues with salmonella, contamination is a far greater issue with factory-farm birds that produce eggs for grocery stores than with your backyard pets. Because salmonella is a foodborne illness, it isn't passed from person to person or hen to hen like a cold. When chickens get salmonella, it is usually the result of eating rat droppings (or worse!) in their tiny, dirty spaces at commercial egg farms. Yuck. (If you buy chicken meat, make sure you cook it thoroughly.)

Salmonella outbreaks at factory farms are another reason to keep your own hens. Presuming you treat your hens humanely, that their coop is clean and free of pests, and that they have fresh food and water at all times, it is doubtful your home flock would contract salmonella. Birds in factory farms have immune systems that are already stressed by the terrible conditions they experience every day.

Talking about Losses with Your Children

With very young children, talking about losses can be difficult, both for you and for them. Naturally, what you do will depend on your child's maturity level and your own personal beliefs. If you're talking about losses of baby chicks, you have some options. For instance, some people prefer not to tell their children exactly how many chicks are expected to arrive, and they open the package privately, away from the children, in case there are any deaths.

The other way to handle the situation is to talk to your young kids ahead of time about the possibility of losses, and if losses do occur, you can deal with them together. Honesty is usually the best policy. Answer questions. Allow them to feel sad about the deaths and sympathize. It's okay for you to be sad, too—it would be terrible if you weren't!—but you have to be sure to model mature behavior for your kids. Allowing your emotions to get out of control will set your kids up to be unable to handle death in a mature way, too.

Either way—whether you decide to discuss it or not—make sure you've already thought through how you plan to handle chicken deaths with your family. And keep in mind that chickens are not always long-lived, and you will have to deal with death at some point in the future.

As your flock ages, you won't have the option to just conceal any losses from your kids. If your hens are your pets, chances are good that your child will notice that Lady Gaga the Polish hen is missing, or that Fluffernutters the Silkie is not moving. Handle the loss the way you would with any pet. Many families keep chickens precisely to teach their children about Mother Nature and the natural cycles of creation and destruction, life and death. Sometimes the loss of a chicken can be a teaching opportunity in disguise.

Despite the minimal risk to your backyard flock, you should still take steps to prevent salmonella illness: Keep the coop clean and your hens happy. Offer them treats containing whole seeds, which research has shown to reduce colonies of salmonella in the gut. And, most importantly, be alert to signs of illness so that if there is a problem, you can take care of it promptly.

A hen that is sick with salmonella will be weak, purple-combed, and have watery diarrhea as well as reduced egg production. If you are worried that your flock has somehow contracted this illness, have them examined by a vet or tested (your local cooperative extension office may offer testing services). But, in most cases, the eggs from your own backyard flock are probably the safest eggs you can eat. When you keep your own birds, you can personally monitor their health, and you can control what feed they eat and the conditions in which they live. You can see when they may need medical attention, and you can provide it.

What about Bird Flu?

Just like salmonella, bird flu is probably not a condition to be excessively worried about—but it is something you should be informed about.

Bird flu, or avian influenza (AI), is a viral disease that can infect domestic poultry. There are two types of AI viruses: low pathogenicity (LPAI) and high pathogenicity (HPAI). Whether a virus is LPAI or HPAI depends on the severity of the illness it causes. HPAI is the extremely infectious and deadly form of the disease that you've heard about in the news. It can spread rapidly from flock to flock. However, there is currently no HPAI virus in the United States, and no human illness from AI has ever occurred here.

Chickens and other domestic birds can get infected with AI through direct contact with infected waterfowl, other infected poultry, and water or feed that has been contaminated—basically, the same way chickens can get other illnesses. But getting LPAI is usually not a great cause for concern, unless your birds are weak or ill with something else. That said, some strains of LPAI are capable of mutating under field conditions into HPAI, meaning they can change from something that is more difficult for chickens to catch into something that is easy for them to catch—and more severe! It is this possibility that the USDA Animal and Plant Health Inspection Service works to prevent.

For instance, HPAI has been detected three

Egg-cetera!

There is no age at which chickens simply stop laying, despite common belief. However, they will lay fewer eggs as they get older. Most laying breeds will lay productively in backyard terms for 5 to 7 years, and sporadically for many years after that.

times in US poultry in 1924, 1983, and 2004. The most recent outbreak was in Texas, but it was quickly eradicated and hasn't been detected here since. As mentioned earlier, no human illness resulted from any of these outbreaks.

Dealing with Death

It should come as no surprise that eventually your chickens will die, due to illness, accident, a predator, or just old age. Losing a pet is always difficult, and losing a pet chicken is no exception. Especially if you have children, you'll want to have a funeral for your pet chicken and

bury her as you would any other pet. The funeral of a pet may be your children's first experience with death, and it can be a learning experience. You may decide to have flowers, a memorial marker, music, a eulogy, the works. Having a funeral for your pet—whether it's a chicken or some other beloved animal—can help your children understand how funerals work, and also can help them let go and find closure.

As you're making your plans, though, check to see what is permissible in your area. How to dispose of a dead pet can vary depending on your state, city, and even your neighborhood—

Losing a favorite bird can be particularly difficult for little ones.

and how the body is handled may depend on the cause of death, too. If your bird displayed signs of illness or died suddenly, for no apparent reason, you'll want to investigate. Either contact your veterinarian, or call the USDA Veterinary Services office for free testing. In many places, if you've lost your bird due to old age or a predator attack, burial is common. Other places may have restrictions, so be sure to check. Some veterinarians can dispose of your dead pet for you, while others may not offer this service—but your local sanitation department might. There are some places that even offer pet cremation. Find out what your local requirements are, and then decide if a pet funeral—and how much of one—will be right for your family.

The best way to discover your own local requirements would be to contact a local vet and ask about them. If no vet is available in your area, you can try contacting your state vet to see what the regulations are for your state, and then contacting local animal control in your county, town, or city to see if there are additional requirements in your area. If you live in a place with a neighborhood association, you will want to check to see if there are any further restrictions on how you must handle the remains of your chicken.

BEHAVIOR and MISBEHAVIOR

Just like with health concerns, there may be a steep learning curve with respect to understanding ordinary chicken behavior during your first year. How do you recognize "ordinary" without any experience, after all?

In this chapter, you'll find tips to help you over the hump, plus a detailed plan of action for one of the biggest challenges inexperienced chicken keepers face: introducing newcomers to the flock.

THE PECKING ORDER

We've talked about pecking order a few times already, and though you probably have a general idea of what it means—heck, it's a fairly common expression, like so many things "chicken" are—it might be helpful at this point to discuss the topic a little more deeply and explain why it has such an effect on introducing new birds to your flock.

Think of the pecking order as a line—a line that your chickens must wait in for services and privileges. In a line, someone is in front and will get served first, while someone else will be at the end and will get served last. But with chickens, it's always the same line, and everyone in line is always in the same order. The hen at the top of the pecking order eats first, gets first dibs on roosting space, and gets first shot at treats. And this is not necessarily a bad thing: If there were no lines, everyone would be scuffling for service at once, pressing their way up to the counter, perhaps trampling each other along the way.

Without a pecking order, every day would be like a particularly vicious Black Friday sale. For chickens.

So, presuming they have plenty of space in the coop, on the roosts, in the yard, and at the feeders and waterers, being at the bottom of the pecking order is not some huge, personal tragedy for chickens, since those at the bottom still have plenty to eat and drink. Chickens take comfort from having their lives ordered this way; it is their instinct, and it is unavoidable. Problems only really arise when there isn't enough for everyone, because then the girls at the end of the line don't get their fair turn.

HOW TO INTRODUCE NEWCOMERS TO YOUR FLOCK

Adding new birds to your established flock can be stressful, both for you and the birds. You may think that because your established flock is peaceful, and because the birds you're adding are peaceful, adding more chickens to your flock is simply a matter of, well, putting them all into the henhouse together. But woe betide to the backyarder who tries that!

Your flock's pecking order will be flung off-kilter by newcomers. Every new bird will have to stake a claim for the primo nesting and roosting spots, and all of your old birds will have to defend their places from the up-and-coming. The new girls won't know that Flora gets dibs on the corner nest, or that Prissy always gets to eat first. Your new birds don't know the flock's rules of civil chicken discourse and will be making all sorts of horrifying social faux pas, like sunbathing in someone else's favorite spot or leaving the coop before Betty does. The horror! It will be as bad as the worst soap opera or middle-school drama you can think of. Fashion will be made

fun of—the ladies with beards or crests, if they are in the minority, will be laughed at relentlessly (or more likely excluded from the lunch group). Or perhaps it will be those without fancy feathering who will be ostracized. There will be bullying, alliances made and discarded, friends won and lost, and so on. Someone will always be in high dudgeon.

War is hell, and so is middle school—and chicken flocks in upheaval.

The good news is that if you handle things correctly, the worst of the pecking order wars should last only a week or two. Plus, there are a few steps you can take to make the introduction process much easier on all of you.

First, make sure the chickens you're introducing are old enough to defend themselves against the larger birds in your flock, certainly no younger than 6 weeks old. Otherwise, when the war breaks out, you could have serious casualties, and even deaths.

Second, and most important, allow the birds to see one another without having actual physical contact for a week or two. This way, they won't be able to hurt each other while establishing dominance, and they'll be able to work out the initial pecking order issues through subtler cues. Pecking at one another without physical contact, through wire mesh, for example, is usually enough to communicate intent and establish some tentative positions in the pecking order, without the danger of anyone getting hurt.

If you have a large coop, you can put your new birds in alongside your old-timers, but separated by chicken wire, a wire dog cage, or something else that'll do the job (mind you, *after* the quarantine period discussed on page 165). Make sure, of course, that all your chickens have access to food and water.

Some people like to remove the barrier at night when the birds have settled down. The idea is that when they wake up, they'll be groggy and more interested in eating than whether or not the barrier is gone, so they may not immediately notice. Others prefer to remove the barrier during the day when the older hens will be outside, so the younger girls will have time to familiarize themselves with the space and explore a little. Both ways work. Best of all, allow your flocks to free range together. They will work out the pecking order but be far too interested in exploring to wage war. After all, there will be bugs to chase and eat. Shady bushes to lounge beneath. Dust holes to scratch out and bathe in. Weeds to find. Feathers to preen. Nests to use. Places to perch. Laps to sit in. Treats to beg. This will last until it's time to go to bed, and, at that point, their priority will be settling down, not starting a fight.

Do expect that despite your precautions, there may still be some jockeying for position. *That's okay.* It's normal. Just monitor the situation as it develops, and keep an eye out to make sure no one is seriously hurt.

If you can't offer lots of space—or even if you can—other distractions can help.

• Hang a half a head of cabbage—or a suet cage with treats, scraps, and greens—barely within reach so the chickens have to work to get at them.

• Temporarily add large branches to the run

and the coop, making pursuit more difficult and giving the new birds places to hide.

• Put straw, dead leaves, mulched dry grass clippings, and/or pulled weeds in their run, giving them plenty to dig through.

You can also employ these same techniques if you're dealing with boredom problems during the winter because your hens are staying indoors more often.

7 CLASSIC MISTAKES TO AVOID WHEN ADDING TO YOUR FLOCK

These tips will help ensure a successful integration of new birds with old.

1. **Don't forget to quarantine the new birds for at least 4 weeks first.** You'll need this time to make sure they don't have any infections or communicable health issues that the rest of your flock can catch. Read more about biosecurity on page 165.

2. **Don't commingle your new birds too soon.** It takes at least a week for the newcomers to establish their spots in the pecking order from within their separate enclosure. Two weeks is preferable.

3. **Don't add just one hen to a large flock.** If you do that, all the flock's aggression and worry will be directed at the one lone, new hen, and it will be difficult to get her integrated successfully. Ideally, add a few hens or more at a time. Even when your flock is

This White Silkie hen would stand out like a sore thumb among an otherwise homogeneous flock.

small—say, just three hens—it will be easier to add two rather than one.

4. **Don't acquire breeds that won't work well with your flock.** If you have a basically homogeneous flock consisting of Rhode Island Reds and other traditional-looking chickens, adding crested Polish or fancy-feathered Faverolles can make for an extraordinarily rough transition period for the new girls. Your established flock may not exactly recognize the newcomers as chickens, much less flock members. If you want to keep many different fancy breeds together, start out with a flock of a few different breeds, or else make the introductions *extra* slow.

5. **Don't introduce chicks when they're too young and too small.** You want to introduce them when they're old enough to take some moderate scuffling, so they don't get hurt.

6. **Don't squeeze too many birds into your space.** Believe us; we do understand that chickens can be addictive. Is it Chickenosis? Chicken Fever? Regardless, there are so many beautiful breeds, you may find yourself wanting one of each, even when you really don't have the infrastructure. So, as you add to your flock, make sure to expand the coop and run space, and add additional nest boxes, feeders, and waterers as necessary. And when you do that: Don't forget that even if you have plenty of feeder/waterer space, adding birds means your feeders and waterers will empty more quickly, so you will need to check and refill them more often.

7. **Don't make it impossible for your new birds to access food and water.** If you're adding younger birds or bantams, remember that your feeders and waterers may need to be lowered so that your new birds can easily reach them.

7 "MISBEHAVIORS" NOT TO WORRY ABOUT

As we've discussed, some normal chicken behaviors are often mistaken for illness by new chicken keepers, and, in Chapter 11, we shared a list of "symptoms" that you *shouldn't* worry about. Likewise, some normal chicken behaviors can be mistaken for misbehaviors—and these shouldn't worry you, either.

1. **Hens being pecked on the head by the rooster.** Presuming no injury is occurring, this is just normal mating behavior. He gives her a peck on the back of the head to express his "romantic" intentions, and she will squat down to be ready for breeding. It doesn't seem as nice as chocolates and flowers, we'll grant you that. But chickens aren't humans. Try to think of it as the way they flirt.

2. **Hens being pecked on the head by another hen.** Between hens, this behavior isn't a prelude to procreation, obviously. But you can think of it as affection, if it helps. It's the establishment and/or reinforcement of the flock's social order. A hen higher in the pecking order has the right to peck anyone lower down than she is; hens lower in the pecking order must submit to a dominant hen's demands or risk being pecked. The occasional peck doesn't amount to much presuming there is plenty of space, food, and water for everyone. Remember, your flock is always in a sort of invisible line. The order in which they wait seldom changes. If a hen tries to cut ahead, she'll get called out for it with a peck on the head. And she'll react either by squatting and demonstrating submission, or by backing off of the behavior in question (moving out of the best spot in the roost or the favorite nest).

3. **Hens squatting down when you pet them.** They're being polite and demonstrating that you're dominant over them. This is how they would react to a peck on the head from a rooster.

4. **Fighting between roosters.** It's not likely in a small backyard flock that you'll want to have more than one rooster. Normally you'll want to maintain a ratio of about 10 hens for every rooster to reduce the chances of overmating the hens and too much competition between the roosters. With most docile backyard chicken breeds—and with sufficient space—a little scuffling is normal and won't be a problem. It will consist chiefly of a few chest bumps and the occasional gotcha-when-you're-not-looking pounce. This substitutes for the head-pecking between hens. But keep

Egg-cetera!

Eggshells with wrinkles or ridges are called body-checked eggs, and result from the hen's body attempting to repair any eggshell damage caused by stress when the egg was inside her. The stress is often caused by disturbances like predators or thunderstorms, or anything that might scare your flock.

an eye out, and if any injuries do occur, take action immediately. You may need to separate the parties or rehome one of them.

5. **Fighting between hens.** Sometimes hens will chest bump, too. It is normal for them to peck at one another and squabble, and less common for them to chest bump, but it does occur—chiefly before they've reached laying age. Hens will even gang up and mount one another. Again, if injuries occur—or if someone is being deprived of food and water for the long term—take action immediately.

6. **A rooster lowering one wing and dancing around a hen.** This courtly behavior is another mating ritual. It's foreplay. When it occurs, it normally precedes the peck on the head. The rooster mesmerizes the hen with his dance moves, strutting and showing off like a bullfighter, sweeping his wings in a most impressive manner. When the hen is sufficiently captivated, she may assume the squatting position on her own— or not. He may have to move on to the head-peck, which is more of a direct proposition.

7. **A rooster dancing around another rooster.** Just like hen-to-hen head pecks, this isn't a prelude to breeding. Instead, the dancing rooster is showing off— demonstrating how very masculine and virile he is! (And beautiful!) He's trying to establish dominance over the other rooster.

Reward Time!

One of the best parts of keeping chickens is getting to enjoy their eggs. So delicious! So fresh! You won't believe the difference, and we warn you now that it may be difficult to enjoy regular eggs again when you go out to eat. They will seem pale, flat, and unappetizing by comparison.

In this section, you'll learn how to gather, store, wash, and—most importantly—cook with your eggs. You'll also discover a wide array of recipes featuring your freshly gathered eggs. And you can show your appreciation to your hens by offering the healthy treat recipes we've included for them, as well.

ENJOYING YOUR HOME-PRODUCED EGGS

In this chapter, you'll find tips on gathering, storing, and cleaning eggs.
There's also information on translating recipe directions to suit
the different sizes of eggs your chickens may be laying.

GATHERING EGGS

Gathering eggs is easy. Simply go out at least once a day to get what your hens have laid. The longer the eggs sit in the nest, the more likely it is that they will accidentally get broken. Your hens may also get broody if many eggs are left in the nest, and that's usually something you want to avoid. Even if you want to encourage broodiness for the purposes of hatching, it's usually better to use golf balls or fake wooden or ceramic eggs for that. Once your hen is broody, then you can give her real (fresh) eggs to hatch. Older eggs hatch with less success.

For the most part—that is, unless they are broody—your hens won't care about their eggs after they have been laid. If you do have a broody girl, and you don't want to hatch eggs, be aware that you may get pecked as you gather eggs from beneath her; you may want to wear long sleeves if that's the case.

You can certainly gather eggs more than once a day, too. Most people who keep backyard flocks gather eggs a couple of times a day, and when you first start getting eggs, you may find it difficult to wait—you may want to keep checking just in case you got another one! You may feel the urge to call everyone you know—or post on social media—to share that you've gotten your first egg, or to tell everyone how many you got today.

A hidden danger of keeping pet chickens: the daily agony of deciding which eggs to crack for breakfast!

WASHING

To wash eggs or not to wash eggs; that is the question! If the eggs are for your own consumption, whether you wash them or not is up to you. But you'll want to be aware of a few facts before you make your decision.

First, washing eggs can be on the risky side. Negligent egg washing tends to be more dangerous than not washing eggs at all, because a wet eggshell provides a great medium for any bacteria on the outside of the eggs to get inside, where they will start growing.

Plus—commercially, at least—there are all sorts of rules attached to washing procedures in an attempt to reduce the risk of contamination that egg washing adds. Regulations in different states may vary, but for instance, the water generally has to be a certain number of degrees warmer than the egg, since cooler water can contribute to the bacteria penetrating the shell. Chemical sanitizer is also used. Finally, the eggs must be thoroughly dried after washing because bacteria can't penetrate a shell that's intact and dry.

It's not just the moisture that causes the potential contamination problem, though. It's the actual washing: Eggs are laid with a natural, protective "bloom" on them—they're designed to be hatched, after all—and this coating helps keep bacteria out so that if the egg is incubated, the chick will be healthy. However, washing removes that natural bloom, in addition to contributing to the risk of bacterial contamination through wet shells.

Another compelling argument for unwashed eggs is the study that *Mother Earth News* did in 1977 that confirmed that eggs keep longer when unwashed; you'll find it at www.motherearth news.com.

So what's the deal? Why would anyone want to wash their eggs?

Simple: It's the law. At least it is in the United States, where eggs from large commercial farms are required by the USDA to be washed with a detergent and then sanitized chemically before they are sold. In the United States, commercial egg producers claim that washing eggs is "necessary for cleanliness." However, washing is only necessary for cleanliness when the eggs aren't produced in clean, humane conditions—in other words, when you're getting your eggs from American factory farms.

In the United Kingdom and Europe, where

Commercial farms withhold funky-looking eggs, but with your backyard flock, you'll find out what you've been missing: There are the occasional tiny "fairy" eggs; eggs with strange lumps and calcium deposits; and extra-large eggs that have no hope of fitting in a traditional egg carton; and, of course, a range of vibrant egg colors!

commercial conditions are often better for chickens, eggs that have been washed cannot be sold as top-grade (Class A) eggs. Class A eggs in these areas must be *unwashed*. This is sensible for a number of reasons. Perhaps the most important is that, since no one wants to buy dirty eggs, there is an additional incentive for the farmers there to provide clean conditions for their layers, when compassion and human decency aren't motivation enough.

So, what's best in the home flock, where hens are properly cared for and have a clean, dry run as well as clean, dry bedding in the coop? We'll say it: Washing eggs that are already clean—and protected with the bloom—is silly. If you sell eggs from your hens, of course be sure to follow your local regulations, silly or not.

STORING

It's a good idea to refrigerate your backyard eggs, but this is also a personal call, presuming your eggs are destined for your own family's consumption only. (Always follow local regulations for washing and storing eggs that are destined for sale to the public.)

In the United States, grocery stores are required to refrigerate eggs; eggs stay fresher longer when refrigerated, since evaporation through the eggshell is reduced in cooler temperatures. In addition, if there is any bacterial contamination inside the egg, it will grow far more slowly in refrigerator temperatures than at room temperature.

By contrast, in the United Kingdom, eggs in grocery stores are kept at room temperature. It seems odd, doesn't it? But the idea is to make sure the eggs stay dry. In a commercial situation where the eggs are being produced at a farm, shipped via trucks, stored at the market, sold to a customer, and then driven home (whew!), repeatedly moving the eggs in and out of refrigeration would cause the shells to get wet with condensation. As we discussed earlier, a wet shell is not a good thing—it provides a vehicle for bacteria to grow.

At home, with eggs you intend to eat, it's sensible to refrigerate, because they can go directly from your coop to your refrigerator, and they will keep longer when refrigerated. Even so, it's pretty common for small flock owners in the United States not to refrigerate eggs for short-term storage before eating them at home. A lot of people use an egg basket or egg skelter (a countertop storage device for unwashed eggs; you load it from the top and the eggs are dispensed from the bottom in order of freshness) so they can really show off the beauty of their flock's eggs. And, in fact, we have to admit that before we hard cook eggs for making deviled eggs, we generally let them sit out on the counter overnight, or even longer. Obviously, this is not to show them off to any unexpected overnight guests! It's because the extra evaporation that will occur at warmer temperatures can help make the eggs easier to peel.

And, in case you're wondering: Store your eggs large end up.

COOKING WITH BACKYARD EGGS

Cooking with your new eggs is not a problem (if you cook, hahaha!). After all, when you're making scrambled eggs—or stirring eggs into fried rice or making egg drop soup—it really won't matter that much if you use four medium rather than four large eggs.

It's baking where you may have some problems, because your backyard eggs may be many different sizes. In baking, the quantity of each ingredient matters a lot more than it does in cooking. Too much moisture in your cake batter (from too much egg) can cause it to fall or not bake through—or it might finish with an eggier flavor than you wanted. Not enough egg can cause your product to be dry or dense, or even just not rich enough. When it comes to baking, deviating significantly from what the recipe calls for could be a recipe for disaster.

For this reason, if you bake a lot, you'll want to be aware of egg size classifications. Here are commercial egg size classifications in the United States.

1 extra-small egg—1.25 ounces

1 small egg—1.5 ounces

1 medium egg—1.75 ounces

1 large egg—2 ounces

1 extra-large egg—2.25 ounces

1 jumbo egg—2.5 ounces or greater

Remember, these are the commercial size classifications for eggs in the United States, but your backyard eggs can vary even from those classifications. For instance, bantam Seramas may lay eggs that weigh only ¾ ounce or so.

If that's the case, then how does one manage? Luckily, it's not hard. Here are four great tips for cooking with backyard eggs.

1. **Don't worry!** You probably won't have to change the way you cook most of the time. You may have to fix two over-easy extra-small eggs rather than one over-easy jumbo egg, but how easy is that? *Easy.* So, to appropriate some wise advice from a wise

man, "Relax, don't worry, and have a backyard egg."

2. **If you bake, get an egg or kitchen scale.** Remember, too, that a young pullet's first eggs will be on the small side for a while. It takes some time for the eggs to attain the full size that the breed lays in maturity, so your egg size will probably grow in the first year or two. Don't get it in your head that your flock always lays medium eggs just because that's what they laid when they first started.

3. **Be aware that the right number of eggs to use in your recipe isn't always the number of eggs listed in the recipe, because unless otherwise specified, published recipes in the United States refer to large eggs.** This means that if your pound cake recipe calls for 8 eggs and your flock lays extra-small eggs, you'll have to use 13 of your backyard eggs to equal the number of eggs the recipe is calling for. You can do the math easily enough, but here is a table of approximate conversions that could help, too.

4. **Vive la différence! Enjoy the size differences in your flock's backyard eggs.** Sometimes bantam eggs are awesome. There is nothing cuter than tiny deviled eggs, tiny pickled eggs, or tiny hard-cooked eggs. (Okay, there may be something cuter in the world, but you get the idea.) If you're making a dish that features an individual egg, using bantam eggs can really be darling. On the other hand, serving up a fist-size scotch egg, or making a three-egg omelet with only one egg can be fun, too.

EGG QUANTITY COMPARISON

Number of large eggs used in recipe

Use this number of the following	1	2	3	4	5	6	7	8
Extra-Small	2	3	5	6	8	10	11	13
Small	1	3	4	5	7	8	9	11
Medium	1	2	3	5	6	7	8	9
Extra-Large	1	2	3	4	4	5	6	7
Jumbo	1	2	2	3	4	5	6	6

RECIPES for EGG LOVERS

When your pet hens finally start laying, you
can celebrate because you'll be blessed with high-quality eggs
that will improve the flavor of every recipe you make.
It will almost inevitably seem like a "finally!" moment
when you find that first egg. It's so exciting that customers
have even told us that they take to social media like Facebook
or Twitter to share photos of those first eggs as if they're
photos of children or grandchildren.

You may think you'll have more eggs than you know what to do with. However, you may not have the abundance of eggs you thought you would. If your plan was to share extras among neighbors, family, and friends, keep in mind you may find you know more egg-loving people than you realized! But even when it comes to eggs for your own immediate family, you'll probably use a lot more eggs than you did before, just because the eggs are so delicious. Plus, eggs are so versatile: They can be used in breakfasts, sides, mains, and desserts—almost anything—to great effect.

BASICS

Best-Ever Scrambled Eggs

6 eggs

3 tablespoons milk

⅛ teaspoon salt (preferably Himalayan pink)

⅛ teaspoon ground black pepper

1 tablespoon unsalted butter

1. In a medium bowl, beat the eggs, milk, salt, and pepper.

2. In a large skillet over medium-high heat, melt the butter. Add the eggs and cook, stirring occasionally, until slightly thickened. Reduce the heat to medium-low and cook, stirring occasionally, until the eggs are set but still moist. Serve immediately. **Makes 6 servings**

Foolproof Hard-Cooked Eggs

12 eggs

Dash of salt

1. In a large pot, arrange the eggs and add enough water to cover. Add the salt.

2. Bring the water to a full rolling boil over high heat. Immediately remove from the heat and let stand for 20 minutes.

3. Plunge the eggs into an ice bath to stop the cooking. Change the ice and water several times until the eggs are completely cool. Peel the eggs and refrigerate until ready to use. **Makes 12**

Notes: The time that you leave the eggs in the boiled water may vary by the size of the egg. Small eggs may take slightly less time, while large eggs may need more time.

By not boiling the eggs, you will avoid unsightly green rings around the yolks. When the eggs get too hot, the iron in the yolk reacts with the sulfur in the egg white to produce a greenish substance called ferrous sulfide. Overcooking eggs can also make the whites rubbery and tough.

Fresh eggs are harder to peel than older eggs. When hard cooking eggs, be sure to use your oldest eggs. Use your freshest for sunny-side up, over easy, baking, poaching, or sharing. (We usually set a carton out the night before we plan to hard cook them, to age them a little.)

Deviled Eggs, Five Ways

12 eggs, hard-cooked (page 190)

¼ cup whipped salad dressing or mayonnaise

¼ cup hot honey mustard (preferably West's Best Hot Honey Mustard*)

½ teaspoon paprika (preferably Hungarian sweet) + additional for garnish

1. Peel the eggs and slice in half lengthwise. Transfer the yolks to a medium bowl and arrange the egg white halves on a serving platter.

2. Mash the yolks. Stir in the dressing or mayonnaise, mustard, and paprika.

3. Spoon or pipe the yolk mixture into the egg white halves. Garnish each deviled egg with the additional paprika. Makes 24

Spicy: Replace the paprika with chili powder.

Super Spicy: Add ⅛ to ¼ teaspoon ground red pepper.

Smoky: Replace the paprika with ⅛ to ¼ teaspoon chipotle chili powder or ½ teaspoon mesquite seasoning.

Mustardy: Add 1 tablespoon of your favorite mustard.

Hatching Chick Deviled Eggs: Cut a small section off of the wide end of the egg to allow it to stand up and not roll on the plate. Cut the egg crosswise in a zigzag pattern about one-third of the way from the top. Remove the yolks and prepare as described above. Spoon or pipe the yolk mixture into the bottom half of the egg white. Place the top on at an angle. Decorate the yolk to look like a chick's face. Use scallions, olives, or capers for the eyes. Use bacon, carrot, or red bell pepper for the beak.

*Lissa's favorite honey mustard for deviled eggs is West's Best Hot Honey Mustard, produced locally in West Virginia. It's sweet and spicy, making the eggs that much richer and creamier.

Red Pickled Eggs

1 cup cider vinegar

2 tablespoons honey

1 teaspoon dry mustard

1 teaspoon ground allspice

1 teaspoon ground ginger

2 cups sliced cooked red beets

4 eggs, hard-cooked (page 190) and peeled

8 cups torn lettuce

1. In a small nonaluminum saucepan, combine the vinegar, honey, mustard, allspice, and ginger and bring just to a boil. Place the beets in a medium heatproof bowl. Pour the hot pickling mixture over them and set aside to cool.

2. When cool, add the eggs and refrigerate overnight. Shake the bowl occasionally so the eggs will color evenly. Store in the refrigerator.

3. To serve, slice or halve the eggs and arrange on a bed of lettuce with the beets. Makes 4 servings

Dressed-Up Egg Salad

¼ cup hot honey mustard (preferably West's Best Hot Honey Mustard)

2 tablespoons whipped salad dressing or mayonnaise

Splash of white wine vinegar

2 scallions, finely chopped

1 tablespoon chopped fresh dill or 1 teaspoon dried dillweed

2 teaspoons paprika or chili powder

12 eggs, hard-cooked (page 190), peeled, and chopped

In a large bowl, combine the mustard, dressing or mayonnaise, vinegar, scallions, dill, and paprika or chili powder. Add the eggs and mix gently until evenly coated. Makes 12 servings

Variations: Stir in any of the following: ¼ teaspoon red-pepper flakes, ground red pepper, or chipotle chili powder; 1 teaspoon curry powder; 2–3 tablespoons grated onion, finely chopped celery, finely chopped pickle, relish, finely flaked smoked salmon, and/or finely diced prosciutto.

To serve as an appetizer, stuff cherry tomatoes with the egg salad or, for a more elegant twist, stuff nasturtium blossoms with the salad.

Note: When deciding which variations to use, keep in mind that one of the advantages of having your own eggs is that they taste so much better than store bought. You may prefer a very basic egg salad to highlight the flavor of the eggs.

Homemade Sesame Mayonnaise

1 egg yolk

½ teaspoon dry mustard

½ teaspoon salt

¼ teaspoon sugar

2 teaspoons champagne vinegar

1 teaspoon lemon juice

¾ cup light olive oil

¼ cup toasted sesame oil

1. In a large bowl, combine the yolk, mustard, salt, and sugar and whisk until blended.

2. In a small bowl, combine the vinegar and lemon juice. In a glass measuring cup, combine the olive oil and sesame oil.

3. Whisk half of the vinegar mixture into the yolk mixture until well blended. Add ¼ cup of the oil mixture a drop or two at a time, whisking constantly, until the mixture thickens. Gradually add half of the remaining oil mixture in a very slow stream, whisking constantly. Slowly whisk in the remaining vinegar mixture, and then the remaining oil mixture. Makes 1 cup

Classic Mayonnaise: Replace the sesame oil and olive oil with canola oil.

BREAKFAST

Cream Cheese Eggs in Crispy Ham Crust

6 slices deli ham

4 eggs

2 tablespoons milk

1 tablespoon butter

1 small onion, finely chopped

4 cloves garlic, minced

3 tablespoons cream cheese

6 grape or cherry tomatoes, halved, for garnish

1. Preheat the oven to 400°F. Lightly oil a 6-cup muffin pan. Line each cup with 1 slice of ham and set aside. In a small bowl, beat the eggs and milk and set aside.

2. In a small skillet over medium heat, melt the butter. Cook the onion for 4 to 5 minutes, stirring occasionally, or until tender and golden. Add the garlic and cook for 1 minute, or until softened.

3. Evenly spoon the onion mixture and ½ tablespoon of the cream cheese into each cup. Evenly pour the egg mixture into the cups.

4. Bake for 25 minutes, or until the eggs are set and the ham is crisp. Using tongs, carefully remove the cups. Garnish with the tomatoes.

Makes 6 servings

Italian-Style Eggs

4 slices (½" thick) French or Italian bread

2 tablespoons canola oil, divided

¼ cup finely chopped onion

¼ cup finely chopped green bell pepper

6 plum tomatoes, chopped

⅛ teaspoon salt

⅛ teaspoon ground black pepper

4 eggs

1. Coat a large nonstick skillet with cooking spray and heat over medium-high heat. Brush both sides of the bread slices with some of the oil. Cook for 4 minutes, turning once, or until toasted. Wipe the skillet clean. Place 1 slice of toast on 4 plates. Set aside.

2. In the same skillet, heat the remaining oil over medium heat. Cook the onion and bell pepper for 2 minutes, stirring occasionally, or until softened. Stir in the tomatoes, salt, and black pepper. Reduce the heat to medium-low and cook for 5 minutes, or until the tomatoes make a chunky sauce.

3. With a large spoon, create 4 indentations in the sauce. Break 2 of the eggs into 2 custard cups. Gently tip each egg into an indentation in the sauce. Repeat with the remaining 2 eggs. Cover and simmer for 6 to 8 minutes, or until the whites are completely set. Use a large spoon to lift each egg and accompanying sauce onto each plate, either next to or on top of the toast. Spoon any remaining sauce evenly around the eggs. Makes 4 servings

Eggs in Rings

1 tablespoon butter

4 whole large ½"-thick onion rings

4 eggs

¼ teaspoon garlic powder (optional)

¼ teaspoon salt

⅛ teaspoon ground black pepper

In a large skillet over medium heat, melt the butter. Cook the onion rings for 5 to 6 minutes, turning once, or until almost tender. Crack 1 egg into each ring and reduce the heat to medium-low. Sprinkle the eggs with garlic powder (if using), salt, and pepper and cook until desired doneness. Makes 4 servings

Veggie Rings: Replace the onion rings with bell pepper rings or hollowed-out squash slices.

Scrambled Egg Rings: Beat the eggs with 2 tablespoons milk and pour into the onion rings.

Lissa's Huevos Rancheros

2 tablespoons corn oil or vegetable oil

6 corn tortillas (6" diameter)

1 can (14–19 ounces) black beans, rinsed and drained

1 can (15 ounces) corn, drained

1 can (14.5 ounces) diced tomatoes with green chiles (preferably Ro-Tel)

1–2 teaspoons Southwest seasoning (preferably Penzeys' Arizona Dreaming)

6 eggs

Toppings: Shredded Cheddar cheese, salsa, sour cream, diced avocado, chopped onion, chives, ramps, and/or scallions (optional)

1. In a large skillet over medium heat, heat the oil. Cook the tortillas for 4 to 6 minutes, turning once, or until crisp. Set aside.

2. In a medium saucepan, combine the beans, corn, tomatoes, and seasoning. Cook over medium heat for 4 to 5 minutes, stirring occasionally, or until heated through. Meanwhile, poach or fry the eggs.

3. To serve, place each tortilla on a plate. Top with some of the bean mixture and an egg. Garnish the huevos rancheros with the cheese, salsa, sour cream, avocado, onion, chives, ramps, and scallions, if desired. **Makes 6 servings**

Poached Eggs on Tomato-Eggplant Beds

1 small eggplant (3" diameter), peeled and cut into 8 slices (¼" thick)

1 tablespoon olive oil

½ teaspoon salt, divided

½ teaspoon ground black pepper, divided

1 tablespoon white vinegar

8 eggs

2 tomatoes, cut into 8 slices (¼" thick)

¼ teaspoon garlic powder

4 slices ham, halved

3 tablespoons chopped fresh basil

1. Preheat the oven to 425°F. Brush both sides of the eggplant slices with the oil and season with ¼ teaspoon of the salt and ¼ teaspoon of the pepper. Place on a baking sheet in a single layer. Bake for 5 to 8 minutes, or just until tender.

2. Meanwhile, heat a large, deep skillet containing 1" of water to a boil over high heat. Add the vinegar and reduce the heat to low. Break an egg into a custard cup and gently tip the egg into the water. Repeat with the remaining 7 eggs. Cover and simmer, shaking the pan 2 or 3 times, for 3 to 5 minutes for a soft-cooked yolk or until the whites are completely set and the yolks begin to thicken.

3. Arrange the eggplant slices on a platter or plates. Place 1 slice of tomato on top of each eggplant slice and season with ⅛ teaspoon of the salt, ⅛ teaspoon of the pepper, and the garlic powder. Top each with 1 slice of ham.

4. Remove the eggs with a slotted spoon and drain over paper towels. Place on top of the ham.

5. Sprinkle with the basil, the remaining ⅛ teaspoon salt, and the remaining ⅛ teaspoon pepper. **Makes 8 servings**

Eggs Florentine with Pesto

1 teaspoon olive oil

1 package (9 ounces) fresh baby spinach

⅓ cup 0% Greek yogurt

¼ cup prepared pesto

1 teaspoon vinegar

Pinch of salt

4 eggs

2 whole grain English muffins, split and toasted

Freshly ground black pepper

1. In a large nonstick skillet over medium-high heat, heat the oil. Cook the spinach (in batches, if necessary) until wilted. In a small bowl, combine the yogurt and pesto. Stir ¼ cup into the spinach and remove from the heat. Cover to keep warm.

2. Meanwhile, heat a medium saucepan containing 1" of water to a boil over high heat. Add the vinegar and salt and reduce the heat to low. Break an egg into a custard cup and gently tip the egg into the water. Repeat with the remaining 3 eggs. Cover and simmer, shaking the pan 2 or 3 times, for 3 to 5 minutes for a soft-cooked yolk or until the whites are completely set and the yolks begin to thicken.

3. Place 1 English muffin half on 4 warm plates. Spoon one-quarter of the spinach mixture onto each muffin. Remove the eggs with a slotted spoon and drain over paper towels. Place on the spinach. Stir 1 tablespoon of the poaching liquid into the remaining yogurt mixture. Spoon evenly over each egg and grind some pepper over the top. Makes 4 servings

Lissa's Sunny Egg Clouds

4 eggs, separated

2 thin slices deli ham, finely chopped + additional for garnish

¼ cup grated aged Asiago cheese + additional for garnish

¼ cup thinly sliced chives or green part of scallion + additional for garnish

1. Preheat the oven to 450°F. Line a baking sheet with parchment paper. With an electric mixer on high speed, beat the egg whites until stiff peaks form. Fold in the ham, cheese, and chives or scallion. Spoon into 4 mounds on the baking sheet and make a deep well in the center of each.

2. Bake for 2 minutes, or until lightly browned. Add 1 yolk to each well. Bake for 2 to 3 minutes, or until the yolks are just set. Garnish with the additional ham, cheese, and chives or scallion. Makes 4 servings

Pineapple-Stuffed French Toast

8 slices Hawaiian sweet bread

4 tablespoons pineapple cream cheese

4 eggs

1 cup milk

2 tablespoons butter, divided

½ cup maple syrup

1. Arrange the bread in four stacks so that equal-size slices are on top of each other. Evenly spread 1 tablespoon of the cream cheese over 1 slice in each stack, leaving a ¼" border. Top each stack with the remaining slice and press gently to seal.

2. In a large, shallow bowl, combine the eggs and milk, whisking until smooth.

3. In a large nonstick skillet over medium-high heat, melt 1 tablespoon of the butter. Working in 2 batches, quickly dip both sides of each sandwich into the egg mixture (do not soak). Transfer to the skillet and cook for 2 to 3 minutes on each side, or until golden brown. Repeat with the remaining butter and sandwiches. Serve warm with syrup. **Makes 4 servings**

Ham Strata

8 ounces ham, finely chopped

1½ cups shredded Swiss cheese

1 tablespoon spicy brown mustard

8 slices whole wheat sandwich bread, lightly toasted

6 eggs

1¾ cups milk

¼ teaspoon garlic powder

1 tablespoon finely shredded aged Asiago or Romano cheese (optional)

1 teaspoon sesame seeds

1. Preheat the oven to 350°F. In a medium bowl, combine the ham, Swiss, and mustard. Spread the ham mixture evenly on 4 slices of toast and top with the remaining slices to make 4 sandwiches. Cut each sandwich on the diagonal into quarters. Arrange the sandwich quarters in a 9" x 9" baking pan, with the points facing up.

2. In a large bowl, beat the eggs, milk, and garlic powder. Pour over the sandwiches and let stand for 10 minutes. Sprinkle with the Asiago or Romano (if using) and sesame seeds and bake for 45 minutes, or until the eggs are set and the tops are lightly browned. **Makes 8 servings**

Green Breakfast Wraps

2 teaspoons butter

1 small zucchini, thinly sliced

1 teaspoon adobo seasoning

1 egg, lightly beaten

2 tablespoons shredded pepper Jack cheese

2 spinach tortillas (10" diameter)

2 slices avocado

2 tablespoons chopped fresh cilantro

¼ cup chopped scallions or ramps

2 tablespoons tomatillo sauce or green salsa

Sliced pickled jalapeño peppers (optional)

1. In a medium skillet over medium heat, melt the butter. Add the zucchini and sprinkle with the adobo seasoning. Cook, stirring occasionally, for 5 minutes, or until tender. Remove the zucchini from the skillet and keep warm. In the same skillet, cook the egg, stirring occasionally, for 2 minutes, or until the egg is set but still moist.

2. Sprinkle the cheese evenly over each tortilla. Heat in the microwave, oven, or toaster oven just until the cheese is melted. Evenly divide the egg and zucchini between the tortillas and top with the avocado, cilantro, scallions or ramps, tomatillo sauce or salsa, and jalapeño peppers (if using). Fold in the sides, and roll up. **Makes 2 servings**

Dutch Baby Puffed Pancake

3 tablespoons butter

3 eggs

¾ cup milk

¾ cup all-purpose flour

¼ teaspoon vanilla extract

1 cup berries (such as blackberries, raspberries, blueberries, sliced strawberries, or gooseberries) or 1 cup of your favorite prepared pie filling

Confectioners' sugar (optional)

1. Preheat the oven to 450°F. In a 9" pie plate or cast-iron skillet, melt the butter. In a medium bowl, combine the eggs, milk, flour, and vanilla and pour into the pie plate or skillet.

2. Bake for 20 minutes, or until puffed and golden brown. The center should collapse when you remove it from the oven. Fill with the berries or filling and sprinkle with confectioners' sugar, if using. **Makes 4 servings**

Dropped Blueberry Scones with Lemon Curd

Scones

⅔ cup milk

1 egg

2 cups all-purpose flour

¼ cup granulated sugar

1 teaspoon baking soda

1 teaspoon cream of tartar

¾ teaspoon salt

1 dash ground nutmeg

½ cup unsalted butter

1 cup blueberries

Turbinado sugar

Butter or jam (optional)

Lemon Curd

2 eggs + 2 egg yolks

¾ cup granulated sugar

⅔ cup lemon juice

⅓ cup unsalted butter

1 tablespoon lemon peel

¼ teaspoon salt

1. To make the scones: Preheat the oven to 400°F. In a small bowl, combine the milk and egg and set aside. In a medium bowl, combine the flour, granulated sugar, baking soda, cream of tartar, salt, and nutmeg. Using a pastry blender or 2 knives, cut in the butter until the mixture resembles coarse crumbs.

2. Gradually stir in the milk mixture just until the dough holds together. Gently fold in the blueberries. Drop the dough by rounded tablespoons onto an ungreased baking sheet and sprinkle with the turbinado sugar.

3. Bake for 10 to 12 minutes, or until golden brown. Serve with lemon curd, butter, or jam (if using).

4. To make the lemon curd: In a double boiler, combine the eggs, yolks, granulated sugar, lemon juice, butter, lemon peel, and salt. Cook over low heat, stirring constantly, for 8 to 10 minutes, or until the curd thickens. Refrigerate, covered, for several hours before serving. **Makes 16 scones and ⅔ cup lemon curd**

MAINS

West Virginia Quiche-Stuffed Pattypan Squash

4 small pattypan squash (3–4" diameter)

1 tablespoon butter

1 large onion, thinly sliced

¼ cup shredded extra-sharp Cheddar cheese

4 eggs

2 tablespoons milk

1–2 small sprigs fresh thyme, stripped (optional) + additional for garnish

Dash of freshly grated nutmeg

Pinch of salt

Paprika (preferably Hungarian sweet)

1. Preheat the oven to 350°F. Cut the tops off of the squash and remove the seeds and flesh, leaving at least a ¼"-thick shell.

2. In a large skillet over medium heat, melt the butter. Cook the onion, stirring occasionally, for 10 to 15 minutes, or until caramelized. Reserve 4 teaspoons for garnish. Spoon the remaining onion evenly into the squash and add 1 tablespoon of the cheese to each.

3. In a medium bowl, beat the eggs, milk, thyme (if using), nutmeg, and salt. Pour into each squash, filling to just below the edge. Sprinkle with the paprika.

4. Arrange the squash on a baking sheet and bake for 20 minutes, or until the egg mixture is set. Garnish with the reserved onions and thyme, if using. Makes 4 servings

Quick Baked Scotch Eggs

1 pound loose sausage

4 eggs, hard-cooked (page 190) and peeled

1. Preheat the oven to 450°F. Divide the sausage into 4 equal portions and flatten each to form a thin patty. Place 1 egg in the center of 1 patty and wrap the sausage around the egg, sealing to completely enclose. Repeat with the remaining sausage and eggs.

2. Arrange the eggs on a wire rack in a baking pan and bake for 30 minutes, or until the meat is no longer pink. Makes 4 servings

Note: These make a hearty meal when served with Rumbledythumps (page 204), a Scottish dish similar to Colcannon.

Scotch Glazed Meatloaves

Glaze

1 cup ketchup

½ cup firmly packed brown sugar

¼ cup finely chopped onion

2 teaspoons liquid smoke

1 teaspoon chili powder

Meatloaf Mixture

1 pound ground beef

1 egg

½ cup dried bread crumbs

1 small onion, finely chopped

2 tablespoons water

1½ teaspoons chili powder

1 teaspoon garlic powder

1 teaspoon salt

½ teaspoon ground black pepper

6 eggs, hard-cooked (page 190) and peeled

1. To make the glaze: In a small bowl, combine the ketchup, sugar, onion, liquid smoke, and chili powder and set aside.

2. To make the meatloaf mixture: In a large bowl, combine the beef, egg, bread crumbs, onion, water, chili powder, garlic powder, salt, and pepper and set aside.

3. Preheat the oven to 400°F. Divide the meatloaf mixture into 6 equal portions and flatten each to form a thin patty. Place 1 egg in the center of 1 patty and wrap the meatloaf mixture around the egg, sealing to completely enclose. Repeat with the remaining patties and eggs. Dip each into the glaze. Arrange the meatloaves in a 13" x 9" baking pan. Bake for 45 minutes, or until 160°F and the meat is no longer pink. **Makes 6 servings**

Scotch Borders Golden Nest Eggs

2 cups Rumbledythumps (below) or mashed
   potatoes

4 eggs

Chopped ramps or scallions, for garnish

Preheat the oven to 400°F. Butter 4 small ramekins, and evenly divide Rumbledythumps or mashed potatoes among them. Make a well in the center of each ramekin. Crack 1 egg into each ramekin. Bake for 30 minutes, or until the desired doneness. Garnish with ramps or scallions. **Makes 4 servings**

Rumbledythumps

2 tablespoons butter

4 cups thinly sliced cabbage

1 onion, chopped

2 cloves garlic, minced

1 pound potatoes, peeled and coarsely chopped

½ cup shredded sharp Cheddar cheese

1. In a large skillet over medium heat, melt the butter. Cook the cabbage and onion, stirring occasionally, for 15 minutes, or until lightly caramelized. Add the garlic and cook for 1 minute.

2. Meanwhile, in a medium saucepan, add the potatoes and enough water to cover. Bring to a boil over high heat. Reduce the heat to low and simmer, partially covered, for 20 minutes, or until the potatoes are tender.

3. Drain the potatoes and mash. Stir in the cabbage mixture and cheese. **Makes 4 servings**

Sesame Ginger Tuna Salad

¼ cup whipped salad dressing or Homemade
Sesame Mayonnaise (page 193)

1 teaspoon toasted sesame oil (see note)

2 teaspoons curry powder

1 teaspoon grated fresh ginger

1 clove garlic, minced

2 cans (5 ounces each) water-packed tuna, drained

*1 egg, hard-cooked (page 190), peeled,
and chopped*

2 scallions, chopped

1 small rib celery with leaves, finely chopped,
or 2 tablespoons finely chopped cucumber

2–3 teaspoons sesame seeds (optional)

Pinch of red-pepper flakes (optional)

In a large bowl, combine the dressing or mayonnaise, oil (if using), curry powder, ginger, and garlic. Stir in the tuna, egg, scallions, celery or cucumber, sesame seeds (if using), and pepper flakes (if using). **Makes 4 servings**

Note: If using Homemade Sesame Mayonnaise, omit the sesame oil.

Smoked Salmon, Asparagus, and Goat Cheese Scrambler

1 tablespoon butter

8 stalks asparagus, woody bottoms removed, cut
into 1" pieces

Salt to taste

Ground black pepper to taste

8 eggs

2 tablespoons fat-free milk

¼ cup crumbled goat cheese

4 ounces smoked salmon, chopped

1. In a large nonstick skillet over medium heat, melt the butter. Cook the asparagus until just tender. Season with the salt and pepper. In a large bowl, beat the eggs, milk, and a few pinches of salt and pepper. Pour into the skillet and cook, stirring occasionally, until slightly thickened.

2. Reduce the heat to medium-low and cook, stirring occasionally, until the eggs are almost set. Stir in the goat cheese and cook for 1 minute, or until the eggs are set but still moist. Remove from the heat and fold in the smoked salmon. **Makes 4 servings**

Lissa's Versatile Quiche

Pie Crust

1⅓ cups all-purpose flour (preferably Hudson Cream Short Patent)

1 teaspoon garlic powder

1 teaspoon dried rosemary or thyme (optional)

½ cup cold unsalted butter

5 tablespoons ice water

Quiche

8 eggs

½ cup milk

⅛ teaspoon freshly grated nutmeg

1 teaspoon paprika (preferably Hungarian sweet or Spanish smoked) + additional for garnish

½ teaspoon salt (optional)

1 cup cooked broccoli or spinach

1 cup shredded Swiss cheese

1. To make the pie crust: In a large bowl, combine the flour, garlic power, and rosemary or thyme (if using). Using a pastry blender or 2 knives, cut in the butter until the mixture resembles coarse crumbs. Gradually stir in the water until the dough just holds together. Wrap the dough with plastic wrap and chill before rolling, if desired. Roll out the dough on a lightly floured surface from the center to the edges, forming a 12" circle. Press into an 8" or 9" pie plate and trim and crimp the edges.

2. To make the quiche: Preheat the oven to 350°F. In a large bowl, beat the eggs, milk, nutmeg, paprika, and salt (if using). Arrange the broccoli or spinach over the crust and evenly sprinkle the cheese on top. Pour the egg mixture into the crust. Garnish the top of the quiche with the additional paprika.

3. Bake for 50 minutes, or until a knife inserted near the center comes out clean. Let cool slightly before serving.
Makes 8 servings

Note: For the filling, you can choose anything that appeals to your palate. Broccoli and spinach are traditional, but anything goes. Sauté your favorite vegetables or use a frozen blend. Just be sure to cook and drain the vegetables before adding them to the quiche. You can also use cooked sausage, bacon, or ham. Or get really creative and make a "pizza quiche" with pepperoni, olives, diced tomato, and Italian seasoning.

Everyday Soufflé

1 tablespoon dried bread crumbs

1 pound broccoli, cut into florets

3 eggs, separated

1½ cups milk

⅓ cup all-purpose flour

¾ teaspoon dry mustard

1 clove garlic, minced

½ cup grated Romano cheese

3 egg whites

⅛ teaspoon cream of tartar

1. Preheat the oven to 375°F. Coat a 2-quart soufflé dish with cooking spray. Add the bread crumbs and shake to coat.

2. Place a steamer basket in a large pot with 2" of water. Bring to a boil over high heat. Place the broccoli in the basket and steam for 8 minutes, or until very tender. Drain and rinse under cold water. Place on a clean kitchen towel to dry. Finely chop and place in a large bowl.

3. In a small bowl, whisk together the yolks and set aside. In a medium saucepan, whisk together the milk, flour, mustard, and garlic and bring to a boil over medium heat. Reduce the heat and simmer, stirring constantly, for 3 minutes, or until slightly thickened. Remove from the heat. Whisk some of the milk mixture into the yolks. Whisk the yolk mixture into the saucepan. Cook for 2 minutes, or until thick.

4. Pour into the bowl with the broccoli and stir in the cheese. In a large mixing bowl with an electric mixer on high speed, beat all 6 egg whites and the cream of tartar until stiff peaks form, occasionally scraping down the sides of the bowl with a rubber spatula. Stir about one-third of the whites into the broccoli mixture. Fold in the remaining whites. Pour into the soufflé dish. Bake for 30 to 40 minutes, or until puffed and golden. **Makes 4 servings**

Crustless Leek and Goat Cheese Quiche

1 tablespoon canola oil or olive oil

2 leeks, white and light green parts, thinly sliced

2 cups sliced cremini mushrooms

2 cloves garlic, minced

4 eggs

½ cup chopped roasted red pepper

½ cup sour cream

2 ounces semisoft goat cheese

¾ teaspoon paprika (preferably smoked)

Salt to taste

Ground black pepper to taste

1. Preheat the oven to 375°F. Grease a 9" tart pan or pie plate. In a large skillet over medium heat, heat the oil. Add the leeks and mushrooms and cook for 5 minutes, or until softened. Add the garlic and cook for 1 minute. Remove the skillet from the heat and transfer the mixture to a bowl to cool slightly.

2. In a separate bowl, beat the eggs, bell pepper, sour cream, goat cheese, paprika, salt, and black pepper. Add the leek mixture and stir to blend. Transfer to the tart pan or pie plate. Bake for 40 minutes, or until set. Let cool slightly before serving. **Makes 4 servings**

Extra-Cheese Monte Cristo

1 egg

¼ cup milk

Dash of freshly grated nutmeg

4 slices bread

4 ounces shredded Gruyère cheese

2 slices deli ham

2 slices deli turkey

2 tablespoons butter

Berry jam (optional)

In a large, shallow bowl, combine the egg, milk, and nutmeg, whisking until smooth. Set aside. On 2 bread slices, layer the cheese, ham, cheese, turkey, and cheese. Top with the remaining bread slices. Dip both sides of each sandwich into the egg mixture. In a large skillet over medium heat, melt the butter. Cook the sandwiches for 2 to 3 minutes on each side, or until golden brown and the cheese is melted. Cut each sandwich in half and serve with berry jam, if using. Makes 2 servings

Croque Madame

Béchamel

1 tablespoon butter

1 tablespoon finely chopped onion

1 tablespoon all-purpose flour

¼ cup milk

¼ cup vegetable or chicken broth

¼ teaspoon salt

⅛ teaspoon ground black pepper

¼ teaspoon paprika, preferably Spanish smoked (optional)

Croque Madame

4 teaspoons prepared yellow mustard

8 slices whole grain sandwich bread

8 ounces thinly sliced deli ham

4 ounces thinly sliced Swiss cheese

2 tablespoons softened butter

4 eggs

1. To make the béchamel: In a small, heavy saucepan over low heat, melt the butter. Cook the onion for 3 minutes, or until just translucent. Add the flour and cook for 3 minutes, stirring constantly. Gradually stir in the milk and broth and cook, stirring constantly, for 5 minutes, or until thickened. Stir in the salt, pepper, and paprika (if using).

2. To make the Croque Madame: Spread the mustard evenly on 4 bread slices and top evenly with the ham and cheese. Top with the remaining bread slices. Spread both sides of each sandwich with the butter. In a large nonstick skillet over medium heat, cook 2 sandwiches for 2 to 3 minutes on each side, or until golden brown. Repeat with the remaining 2 sandwiches.

3. Preheat the broiler. Arrange the sandwiches on a baking sheet. Spoon the béchamel evenly over the sandwiches and broil a few inches from the heat for 2 to 3 minutes, or until the sauce is slightly browned.

4. Meanwhile, in the same skillet, break 2 eggs and cook over medium heat to desired doneness. Repeat with the remaining 2 eggs. Top the sandwiches with the eggs and serve immediately. Makes 4 servings

Salmon Patties with Cucumber Ribbons and Dynamite Sauce

Salmon Patties

1 pound salmon fillet

2 eggs

1 teaspoon toasted sesame oil

¼ cup finely chopped scallions

¼ cup dried bread crumbs

¼ cup all-purpose baking mix

½ teaspoon ground ginger

½ teaspoon garlic powder

1 tablespoon vegetable oil

1 small cucumber

Dynamite Sauce

¼ cup plain yogurt

¼ teaspoon garlic powder

1 tablespoon Sriracha hot sauce

Splash of teriyaki sauce

1. To make the salmon patties: Preheat the oven to 400°F. Arrange the salmon on a baking sheet and bake for 15 minutes, or until the fish flakes easily. Let the salmon cool and break into pieces. In a large bowl, combine the flaked salmon, eggs, oil, scallions, bread crumbs, baking mix, ginger, and garlic powder. Form into 4 patties. In a large skillet over medium heat, heat the oil. Add the patties and cook for 2 minutes on each side, or until golden.

2. Cut the cucumber in half crosswise. Using a vegetable peeler, peel long strips to form ribbons.

3. To make the Dynamite Sauce: In a small bowl, combine the yogurt, garlic powder, Sriracha, and teriyaki sauce. Top the salmon patties with the Dynamite Sauce and cucumber ribbons.

Makes 4 servings

Provençal Mushroom-and-Vegetable Main Dish Salad

1 teaspoon + 3 tablespoons olive oil

1 cup dried bread crumbs

2 teaspoons herbes de Provence

2 teaspoons paprika

6 large portobello mushrooms

¼ cup Dijon mustard

1 bag (5 ounces) spring salad mix

½ cup drained roasted red bell peppers, sliced

3 eggs, hard-cooked (page 190), peeled, and sliced

1 bunch scallions, white and light green parts, sliced

2 cups trimmed green beans, steamed and patted dry

¾ cup vinaigrette salad dressing

1. Preheat the oven to 400°F. Coat a large baking sheet with 1 teaspoon of the oil. In a large, shallow bowl, combine the bread crumbs, herbes de Provence, and paprika. Coat both sides of the mushrooms evenly with the mustard. One at a time, dip both sides of the mushrooms into the bread crumb mixture. Place in a single layer on the baking sheet. Drizzle with the remaining 3 tablespoons oil.

2. Bake for 20 minutes. Let stand to cool completely. Slice into ¼"-thick strips.

3. Arrange the spring mix on a large serving platter. Top with the mushrooms, peppers, eggs, scallions, and beans. Drizzle the dressing over the salad. **Makes 6 servings**

SIDES

Salad Niçoise

Salad

1 head Bibb or romaine lettuce

2 ripe tomatoes, each cut into 6 wedges

4 red potatoes, cooked and quartered

½ cup green beans, blanched

4 eggs, hard-cooked (page 190), peeled, and quartered

⅛ cup cured black olives

½ cup drained water-packed tuna

8 anchovy fillets

Dijon Mustard Vinaigrette

2 teaspoons Dijon mustard

1 tablespoon red wine vinegar

Salt to taste

Ground black pepper to taste

1 cup olive oil

1. To make the salad: Arrange the lettuce leaves on the bottom of a bowl or serving platter. Arrange the tomatoes, potatoes, green beans, eggs, and olives over the lettuce. In the center, place the tuna and top it with the anchovies.

2. To make the Dijon mustard vinaigrette: In a small bowl, combine the mustard, vinegar, salt, and pepper. Slowly whisk in the oil. Drizzle the dressing over the salad. Makes 4 servings

Egg Noodles

2 cups all-purpose flour (preferably Hudson Cream Short Patent)

½ teaspoon salt

¼ teaspoon baking powder

5 egg yolks

½–¾ cup milk

1. In a large bowl, combine the flour, salt, and baking powder. Add the yolks and mix until the dry ingredients are moistened. Gradually stir in just enough of the milk until the mixture forms into a ball.

2. Divide the dough into 4 equal parts. Roll out on a floured surface, one part at a time, into a rectangle about ⅛ to ¼" thick. Cut the noodles to the desired width and length. Lay on a linen dish towel or wooden dowel for 30 minutes to 2 hours to dry.

3. Fill a large pot with water and 2 teaspoons salt or stock and bring to a boil. Add the noodles and cook for 1 to 3 minutes, or until the noodles float. **Makes 1¼ pounds noodles**

Quick Egg Drop Soup

4 cups vegetable or chicken broth

1 teaspoon grated fresh ginger or ½ teaspoon ground

Pinch of Chinese five-spice powder (optional)

3 eggs, beaten

½ cup fresh baby spinach, coarsely cut

1 small ripe tomato, chopped (or a handful of halved grape tomatoes)

2–3 scallions, finely chopped with tops

Dash of toasted sesame oil

Dash of soy sauce

1. In a medium saucepan, combine the broth, ginger, and five-spice powder (if using) and bring to a boil over high heat. Reduce the heat to a simmer. Slowly pour in the eggs, while constantly stirring the soup until long strands of cooked egg are formed.

2. Remove from the heat and add the spinach, tomato, scallions, oil, and soy sauce. Let stand for 4 minutes, or until the spinach is wilted and the tomato and scallions are heated through. **Makes 4 servings**

Grandma Waterman's Potato Salad

2 pounds potatoes (preferably red), peeled if desired

1 teaspoon salt

4 eggs, hard-cooked (page 190), peeled, and chopped

1 cup chopped celery

½ cup chopped onion

½ cup minced fresh parsley

¼ cup French dressing (preferably Catalina or California)

¾ cup whipped salad dressing or mayonnaise

1 tablespoon prepared yellow mustard

4 teaspoons lemon juice

2 teaspoons celery seed

Salt to taste

Ground black pepper to taste

1. In a 4-quart saucepan, add the potatoes, enough water to cover, and the salt. Bring to a boil over high heat. Reduce the heat to low and simmer, partially covered, for 20 minutes, or until the potatoes are tender. Drain, let cool, and dice.

2. In a large bowl, combine the potatoes, eggs, celery, onion, parsley, and French dressing. Refrigerate several hours or overnight, allowing the flavors to blend.

3. In a small bowl, combine the salad dressing or mayonnaise, mustard, lemon juice, celery seed, salt, and pepper. Add to the potato mixture and toss gently until evenly coated. **Makes 8 servings**

Fried Brown Rice

1 cup brown rice

Pinch of salt

3 tablespoons peanut oil or vegetable oil

2 cups chopped Chinese cabbage

¾ cup snow peas, snap peas, and/or broccoli florets

1 tablespoon chopped Cuban seasoning pepper or red bell pepper

2–3 tablespoons soy sauce (or to taste)

3 eggs, beaten

1 bunch scallions, sliced

1 teaspoon toasted sesame oil

1. Cook the rice according to package directions, adding the salt.

2. Meanwhile, in a large wok or skillet over medium-high heat, heat the oil. Cook the cabbage; snow peas, snap peas, and/or broccoli; and pepper, stirring frequently, until tender-crisp. Reduce the heat to medium and add the cooked rice and soy sauce. Cook for 3 minutes, stirring frequently, or until the rice is heated through. Push the rice mixture to one side of the wok and pour the eggs into the other side. Cook the eggs, stirring occasionally, until set. Stir the eggs into the rice mixture. Stir in the scallions and sesame oil and serve immediately. **Makes 4 servings**

Potato and Onion Frittata

2 baking potatoes, peeled, halved lengthwise, and thinly sliced crosswise

1 tablespoon olive oil

1 onion, thinly sliced

6 eggs

¼ teaspoon curry powder

¼ teaspoon ground ginger

1. Place the oven rack so it is 6" from the broiler and preheat the broiler to high. Steam the potatoes for 5 minutes, or until tender. In a large ovenproof skillet over medium heat, heat the oil. Cook the onion for 5 minutes, or until translucent. Add the potatoes and spread the mixture so it forms an even layer on the bottom of the skillet.

2. In a medium bowl, whisk the eggs, curry powder, and ginger until blended. Pour into the skillet and cook for 5 minutes, or until the edges are set. Transfer to the oven and broil for 4 minutes, or until the frittata is puffy around the edges and golden. **Makes 4 servings**

Zucchini alla Napoli

2 small zucchini, grated

1 egg

½–¾ cup all-purpose baking mix

2 tablespoons dried bread crumbs

2 tablespoons grated aged Asiago cheese + additional for garnish

1 tablespoon Italian seasoning

1 teaspoon garlic powder

2 tablespoons olive oil, divided

1 cup marinara sauce, heated

1. Press the grated zucchini with a paper towel to remove excess moisture. In a large bowl, combine the zucchini, egg, ½ cup of the baking mix, the bread crumbs, cheese, Italian seasoning, and garlic powder. Gradually add enough of the remaining baking mix until the mixture is the consistency of very thick pancake batter.

2. In a large skillet over medium heat, heat 1 tablespoon of the oil. Drop heaping tablespoons of the batter into the skillet. (You can use more or less batter depending on the desired size of the patties.) Cook the pancakes for about 2 minutes on each side, or until browned. Place the pancakes on a sheet pan and keep warm in the oven. Repeat with the remaining oil and batter until all of the batter is used.

3. To serve, arrange the pancakes on a serving platter or plates. Top with the marinara sauce and garnish with the additional cheese. Serve with garlic bread, if desired. **Makes eight 3" pancakes**

Variation: You can replace 1 tablespoon Italian seasoning with 1½ teaspoons oregano, ½ teaspoon marjoram, ½ teaspoon basil, a pinch of red-pepper flakes, and a pinch of fennel.

Egg Foo Yung

2 cups thinly sliced or shredded cabbage
or bok choy

¾ cup thinly sliced scallions

¾ cup shredded carrots

¼ cup mung bean or alfalfa sprouts

1 teaspoon soy sauce

¼ teaspoon salt

⅛ teaspoon Chinese five-spice powder

5 eggs

1 tablespoon grated fresh ginger

2 tablespoons water

6 teaspoons ground flaxseeds

1. Coat a large nonstick skillet with cooking spray and heat over medium heat. Cook the cabbage or bok choy, scallions, carrots, sprouts, soy sauce, salt, and five-spice powder, covered, for 7 to 10 minutes, stirring occasionally, or until the vegetables are wilted and lightly browned.

2. Meanwhile, in a medium bowl, beat the eggs, ginger, and water with a fork. Coat a 9" nonstick omelet pan with cooking spray and heat over medium heat. Ladle one-quarter of the egg mixture (5 tablespoonfuls) into the pan. Cook for 20 to 30 seconds, or until the edges start to set. Using a silicone spatula, carefully lift the edges, tipping the pan to allow the runny mixture to get underneath. When the eggs are almost set and just shimmering on top, 1 minute, sprinkle on 1½ teaspoons of the flaxseeds and ½ cup of the vegetable mixture. Cook for 30 seconds, or until the eggs are completely set. Slide the pancake onto a dinner plate or roll the pancake like a jelly roll before sliding onto the plate. Repeat for the remaining 3 pancakes. **Makes 4 servings**

Corn Custard

1 cup evaporated milk

4 eggs

2 tablespoons chopped onion

½ teaspoon salt

½ teaspoon ground black pepper

4 cups fresh or frozen and thawed whole kernel corn, divided

Preheat the oven to 325°F. Coat an 8" x 8" baking dish with cooking spray. In a blender or food processor, combine the milk, eggs, onion, salt, pepper, and 2 cups of the corn and puree until smooth. Transfer to a large bowl. Stir in the remaining 2 cups corn. Pour into the baking dish. Place the baking dish in a larger ovenproof pan. Add hot water to the larger pan to a depth of 1". Bake for 1 hour 15 minutes, or until a knife inserted in the center of the custard comes out clean. **Makes 4 servings**

DESSERTS

Lissa's Solstice Egg Nog

6 eggs, separated

4 cups milk

1 can (12 ounces) evaporated milk

7 ounces (½ of 14-ounce can) sweetened
 condensed milk

1 teaspoon vanilla extract
 (preferably Madagascar bourbon)

Freshly grated nutmeg, for garnish

1. In a large bowl, with an electric mixer on medium speed, beat the yolks until they turn pale yellow. Gradually add the milk, evaporated milk, condensed milk, and vanilla.

2. In another large bowl, with an electric mixer on medium speed, beat the egg whites until stiff peaks form. Using a wire whisk, gently fold the egg whites into the yolk mixture. Chill and garnish with the nutmeg. **Makes 8 cups**

Epiphany Lemon Bars

Crust

1 cup all-purpose flour

¼ cup confectioners' sugar + additional for garnish

½ teaspoon grated lemon peel

½ cup unsalted butter

Filling

2 eggs

¾ cup granulated sugar

⅓ cup lemon juice

2 tablespoons all-purpose flour

¼ teaspoon baking powder

1. To make the crust: Preheat the oven to 350°F. Line a 9" x 9" baking pan with parchment paper. In a large bowl, combine the flour, confectioners' sugar, and peel. Using a pastry blender or 2 knives, cut in the butter until the mixture resembles coarse crumbs. Spread the crust mixture evenly over the bottom of the pan, pressing firmly. Bake for 10 minutes.

2. To make the filling: In a large bowl, with an electric mixer on medium speed, beat the eggs until thick and fluffy. Beat in the granulated sugar and lemon juice. Gently stir in the flour and baking powder just until incorporated. Pour over the crust and bake for 20 to 25 minutes, or until the edges just start to turn golden. Cool completely. Garnish with the additional confectioners' sugar and cut into 3" x 2¼" bars. **Makes 12 bars**

Profiteroles

Cream Puffs

1 cup water

½ cup butter

¼ teaspoon salt

1 cup unbleached all-purpose flour

4 eggs

Pastry Cream

1 tablespoon cornstarch

1 cup milk, divided

1 egg

1 egg yolk

¼ cup granulated sugar

Pinch of salt

2 tablespoons butter

1 teaspoon vanilla extract

Confectioners' sugar, for garnish

1. To make the cream puffs: Preheat the oven to 400°F. In a medium saucepan over high heat, bring the water, butter, and salt to a boil. Stir in the flour until smooth. Reduce the heat to low and cook, stirring constantly, for 30 seconds, or until the mixture forms a ball. Remove from the heat. Beat in the eggs, one at a time, until blended and smooth. Drop by rounded teaspoons about 1½" apart onto ungreased baking sheets. Bake for 22 minutes, or until puffed and golden. Cool completely on a rack.

2. To make the pastry cream: In a medium bowl, whisk together the cornstarch and ¼ cup of the milk until the cornstarch is dissolved. Whisk in the egg and yolk. Set aside. In a medium saucepan, combine the granulated sugar, salt, and the remaining ¾ cup milk. Bring to a boil over medium heat. Slowly whisk about ½ cup of the milk mixture into the egg mixture. Whisk back into the saucepan and cook, whisking constantly, for 2 minutes, or until the mixture thickens. Remove from the heat.

3. Whisk in the butter and vanilla. Pour the pastry cream into a glass bowl and cover with plastic wrap against the surface. Refrigerate for 1 hour, or until cold. Pierce the bottom of each puff with a knife. Spoon the chilled pastry cream into a plastic food storage bag and snip ¼" off a corner of the bag. Pipe the cream into each puff. Arrange on a serving plate and dust with confectioners' sugar. **Makes 4 dozen**

Classic Chocolate Soufflé

1 tablespoon + ½ cup granulated sugar

6 ounces semisweet chocolate, chopped

¼ cup butter

4 eggs, separated

⅓ cup unbleached all-purpose flour

1 tablespoon vanilla extract

Confectioners' sugar, for garnish

1. Preheat the oven to 375°F. Grease six 6-ounce soufflé cups or ramekins and sprinkle with 1 tablespoon of the sugar. Set aside.

2. In a small saucepan over low heat, melt the chocolate and butter. Set aside to cool. In a medium bowl, with an electric mixer on medium speed, beat the egg whites until foamy. Slowly add the remaining ½ cup granulated sugar and beat until stiff peaks form.

3. In a large bowl, whisk together the yolks, flour, and vanilla. Slowly whisk in the chocolate mixture. Stir in one-quarter of the egg white mixture. Gently fold in the remaining egg white mixture.

4. Evenly spoon into the cups. Bake for 20 minutes, or until the soufflés are puffed and just set. Dust with the confectioners' sugar and serve immediately. **Makes 6 servings**

Baked Barley-Cherry Custard

4 cups milk

¼ cup pearled barley

¼ teaspoon salt

2 eggs

¼ cup chopped dried cherries

2 tablespoons brown rice syrup, malt syrup, or honey

½ teaspoon vanilla extract

Ground cinnamon, for garnish

1. In a large saucepan, combine the milk, barley, and salt. Cook over medium-low heat for 45 to 55 minutes, or until the barley is tender.

2. Preheat the oven to 350°F. In a small bowl, beat the eggs. Add the cherries, syrup or honey, and vanilla. Slowly stir the egg mixture into the barley mixture and cook over very low heat for 5 minutes, stirring constantly. Spoon into 6 custard cups and sprinkle lightly with the cinnamon. Bake for 10 to 15 minutes, or until the custard is set. Cool and refrigerate. Serve chilled. **Makes 6 servings**

Crème Brûlée

1 tablespoon butter, melted

2 cups light cream

½ cup granulated sugar

1 vanilla bean, split lengthwise

6 egg yolks

¼ cup firmly packed brown sugar

1. Preheat the oven to 300°F. Butter six 6-ounce custard cups. Place in a baking pan and set aside.

2. In a medium saucepan, combine the cream, granulated sugar, and vanilla bean. Cook over medium heat, stirring occasionally, until the sugar is dissolved and the mixture just comes to a boil. Remove from the heat. Remove the vanilla bean and scrape the seeds back into the cream mixture. Discard the bean.

3. In a medium bowl, whisk the yolks until smooth. While whisking constantly, gradually add the cream mixture until well blended. Evenly pour the custard into the custard cups. Add warm water to the baking pan until it reaches halfway up the sides of the cups. Bake for 40 minutes, or until the custard is set in the center. Remove from the oven and water bath and cool to room temperature. Chill for at least 2 hours or overnight.

4. Place the oven rack 4" from the broiler and preheat the broiler to high. Evenly spoon 2 teaspoons of the brown sugar over each cup. Place on a baking sheet and broil for 1 to 2 minutes, turning every 30 seconds, or until evenly brown. Refrigerate for 30 minutes, or until the custard is chilled and the topping has hardened. *Makes 6 servings*

Pumpkin Custard

1 can (15 ounces) pumpkin puree

2 eggs

1 cup fat-free evaporated milk

1 cup cold water

½ cup sugar

1 teaspoon vanilla extract

½ teaspoon freshly grated nutmeg

In a large saucepan, whisk together the pumpkin, eggs, milk, water, sugar, vanilla, and nutmeg. Bring to a simmer over medium heat, stirring occasionally. Continue simmering, stirring occasionally, until the mixture thickens. Pour the custard into 6 custard cups and chill before serving. *Makes 6 servings*

Bread and Butter Pudding

⅓ cup raisins

¼ cup rum extract or whiskey

2 cups milk

⅓ cup heavy cream

¾ cup peach or apricot fruit spread

2 teaspoons vanilla extract

¾ teaspoon ground cinnamon (preferably cassia)

Pinch of salt

4 eggs, at room temperature

3 egg yolks, at room temperature

8 slices day-old whole wheat bread

3 tablespoons unsalted butter, softened

1 tablespoon confectioners' sugar (optional)

1. Coat an 8" x 8" baking pan with cooking spray. In a microwaveable cup or mug, combine the raisins and rum extract or whiskey. Microwave on high power for 15 to 20 seconds, or until hot. Cover to keep warm. In a large saucepan, combine the milk, cream, fruit spread, vanilla, cinnamon, and salt. Cook over low heat for 8 to 10 minutes, stirring occasionally, or until tiny bubbles appear around the edge of the saucepan.

2. Meanwhile, in a large bowl, whisk the eggs and yolks. Gradually whisk ½ cup of the milk mixture into the eggs. Quickly whisk in the remaining milk mixture. Strain the raisin mixture over the milk-egg mixture, reserving the raisins. Spread 1 side of each bread slice with the butter. Cut the bread into cubes and arrange in the bottom of the baking pan in an even layer. Scatter the raisins over the bread and pour the milk-egg mixture over top. Press a piece of plastic wrap directly onto the surface of the mixture and let sit until the bread is thoroughly soaked, 30 to 40 minutes, occasionally pressing down on the plastic wrap to keep the bread submerged.

3. Preheat the oven to 350°F. Discard the plastic wrap and bake for 45 to 50 minutes, or until a knife inserted in the center comes out clean. Serve warm or at room temperature. Dust with the confectioners' sugar, if using. **Makes 8 servings**

Honey-Vanilla Ice Cream

2 cups cold milk, divided

⅓ cup honey (preferably clover or orange blossom)

1 tablespoon cornstarch

2 egg yolks, beaten

2 cups light cream

1 tablespoon vanilla extract

1. In a small bowl, combine ½ cup of the milk, the honey, and cornstarch. Stir until smooth. In the top of a double boiler set over hot water, heat the remaining 1½ cups milk until just about to boil. Stir in the cornstarch mixture very slowly. Cook for 8 minutes. Add the yolks and cook for 2 minutes. The mixture should be thick and smooth. Remove from the heat and let cool.

2. Stir the cream and vanilla into the mixture. Pour into the canister of an ice cream maker. Freeze according to the manufacturer's instructions. **Makes 12 servings**

Angel Food Cake

1 cup + 2 tablespoons sifted cake flour

1¼ cups + 2 tablespoons sugar, divided

12 egg whites, at room temperature

1¼ teaspoons cream of tartar

¼ teaspoon salt

1½ teaspoons vanilla extract

½ teaspoon almond extract

1. Preheat the oven to 350°F. Sift the flour with ½ cup of the sugar 3 times. Set aside.

2. In a large bowl, combine the egg whites, cream of tartar, and salt. With an electric mixer on medium speed, beat the egg whites until soft peaks form. Gradually beat in the remaining sugar, about ¼ cup at a time. Add the vanilla and almond extract. Beat until glossy.

3. Gently fold the flour mixture into the egg white mixture in 3 parts, folding just enough to mix each time. Gently pour the batter into an ungreased 9" or 10" straight-side nonstick tube pan and smooth the top.

4. Bake for 40 to 45 minutes, or until the cake is golden brown. Invert the pan over the top of a heavy or long-neck bottle or funnel and let cool for 1½ to 2 hours. Gently remove to a serving plate. **Makes 12 servings**

Lemon Poppy Seed Pound Cake

2¼ cups all-purpose flour

2 teaspoons baking powder

¾ teaspoon salt

1 cup unsalted butter, at room temperature

1½ cups sugar

1 tablespoon grated lemon peel

3 eggs, at room temperature

½ cup sour cream

¾ cup milk

½ cup poppy seeds

1. Preheat the oven to 350°F. Butter and flour a 9" x 5" loaf pan. In a medium bowl, combine the flour, baking powder, and salt. Set aside. In a large bowl, with an electric mixer on medium-high speed, beat the butter, sugar, and peel for 3 minutes, or until light and fluffy. Beat in the eggs, one at a time, until light and fluffy. Beat in the sour cream until just incorporated.

2. Gradually add the flour mixture to the butter mixture in 3 parts, alternating with the milk, beginning and ending with the flour mixture. Fold in the poppy seeds by hand. Spoon the batter into the pan and smooth the top.

3. Bake for 55 to 60 minutes, or until a toothpick inserted in the center comes out clean. Let cool in the pan for 10 minutes. Remove from the pan and transfer to a rack to cool completely. Wrap with plastic until ready to serve.

Makes 12 servings

Berries and Cream Sponge Cake

6 eggs, separated

1 tablespoon grated orange peel

½ cup orange juice

½ cup + 2 tablespoons honey, warmed

1⅓ cups whole wheat pastry flour

¼ teaspoon salt

1 teaspoon cream of tartar

1 pint heavy or whipping cream

2 cups fresh berries + additional for garnish

1. Preheat the oven to 325°F. In a large bowl, with an electric mixer on high speed, beat the yolks for 5 minutes, or until thick and pale yellow in color. Add the peel and orange juice and beat for 5 minutes. Gradually beat in ½ cup of the honey, 1 tablespoon at a time, until the mixture is very thick and the batter flows back into the bowl in a thick ribbon when the beaters are lifted, 12 to 15 minutes. Do not underbeat—the lightness of this sponge cake depends on the beating process. Evenly sprinkle the flour over the yolk mixture and gently fold it in. Set aside.

2. In another large bowl, with an electric mixer on high speed, beat the egg whites and salt until foamy. Add the cream of tartar and beat until the mixture forms soft peaks. Gradually add the remaining 2 tablespoons honey and beat until the mixture is stiff but not dry. With a rubber spatula, gently fold into the yolk mixture. Pour the batter into 3 ungreased 8" round cake pans. Bake for 20 to 22 minutes, or until the tops spring back when lightly touched and the cakes are beginning to pull away from the sides of the pans. Cool the cakes completely in the pans. Run a metal spatula around the edges of the pans and invert the cakes onto a rack.

3. In a large bowl, with an electric mixer on medium speed, beat the heavy cream until stiff peaks form. Place 1 cake layer on a serving plate, spread ¾ cup of the whipped cream over the top of the cake, and sprinkle with 1 cup of the berries. Place another cake layer on top and spread another ¾ cup of the whipped cream over the top of the cake. Sprinkle with the remaining 1 cup berries. Place the remaining cake layer on top and frost the top and sides of the cake with the remaining whipped cream. Garnish the top of the cake with the additional berries. **Makes 8 servings**

TREATS for YOUR CHICKENS

Do you need to cook for your chickens? Of course not. In fact, if you choose to make them special treats, you should take care to limit their intake and ensure that they get most of their nutrition from their balanced feed. Even so, everyone likes a treat now and then, chickens included!

You may want to create your own recipes, and that's great. Just keep in mind that there are a few things chickens shouldn't eat, at least in quantity, so you should leave those things out of any recipes you design. On the top of that list are dairy products. That means no milk, no cheese, no ice cream (as if you would!). A little plain live-culture yogurt can be given sparingly, as it has helpful probiotics. But that's about it as far as dairy goes. Mammals are designed to digest milk; birds aren't.

You may also want to leave out the onions and garlic. While these aren't bad for your chickens per se, they can cause your hens' eggs to have an oniony or garlicky taste. If you're making a savory quiche, garlicky eggs might be okay—but they probably wouldn't go over as well in vanilla pudding or pound cake.

In addition, avocados can be bad for birds. Not all birds are sensitive to persin, which is the toxin found in avocado—and some types of avocado (as well as some parts of it) seem to be more toxic than others. Just to be safe, though, skipping avocado is a good idea.

Another thing to consider is that very sweet or very salty foods should also be used sparingly. Too much salt, especially, can lead to issues with thin eggshells. Too much sugar can mean diarrhea or loose stools.

Lastly, beware of potato skins—if they're green. Green potato skins contain solanine, which is bad not only for chickens, but also for humans. To be fair, you'd have to eat an awful lot of green potato skins to experience an adverse effect, but it's just better to be safe than sorry for yourself and your chickens.

Hatchday Cake for Chickens

⅔–1 cup whole wheat flour

½ cup yellow cornmeal

1 tablespoon baking powder

¼ teaspoon salt

1 egg

1 can (15 ounces) cream-style corn

1 can (15.25 ounces) whole kernel corn (undrained)

2 tablespoons canola oil

Plain yogurt or refried beans

Preheat the oven to 350°F. Grease a 1½-quart baking dish. In a large bowl, combine ⅔ cup of the flour, the cornmeal, baking powder, and salt. Stir in the egg, cream-style corn, whole kernel corn with its liquid, and oil. The consistency should be similar to cornbread batter. Add a little more flour if needed to thicken. Pour the batter into the baking dish and bake for 30 to 35 minutes, or until a toothpick inserted in the center comes out almost clean. Frost sparingly with plain yogurt or refried beans.

Hatchday Cupcakes: Grease a 12-cup muffin pan. Pour the batter evenly into the cups. Reduce the baking time to 20 minutes, or until a toothpick inserted in the center comes out almost clean.

Scrambled Eggs for Chickens

4 eggs

2 tablespoons water

1 tablespoon corn oil or canola oil

2 tablespoons finely chopped green bell pepper (optional)

1. In a medium bowl, beat the eggs and water.

2. In a large skillet over medium-high heat, heat the oil. Cook the eggs and pepper (if using), stirring occasionally, until slightly thickened. Reduce the heat to medium-low and cook, stirring occasionally, until the eggs are set but still moist.

Porridge for Chickens

1 cup water

½ cup rolled oats

¼ cup finely chopped apple or raisins (chickens' choice!)

2 tablespoons soy or casein protein powder

2 teaspoons flaxseeds (optional)

In a small saucepan, combine the water, oats, and apple or raisins. Bring to a boil over high heat. Reduce the heat to low and simmer, stirring occasionally, until thickened. Stir in the protein powder and sprinkle with the flaxseeds, if using.

Note: This is a warm treat for a cold morning.

Waldorf Salad for Chickens

1 apple, chopped

1 cup seedless grapes, halved

½ cup hulled pumpkin seeds

½ cup raisins

½–1 cup plain yogurt

In a medium bowl, combine the apple, grapes, seeds, raisins, and yogurt. Mix well.

Note: Serve this on a hot day for a refreshing treat.

Summer Fruit Salad for Chickens

1 ripe banana, chopped

1 apple, chopped

½ cantaloupe and/or honeydew melon, chopped

1 cup halved strawberries and/or grapes

1 cup blackberries and/or raspberries

In a large bowl, combine the banana, apple, melon, strawberries or grapes, and blackberries or raspberries. Refrigerate until ready to serve.

Note: This makes a cool treat for a hot summer day.

Pasta Salad for Chickens

8 ounces whole grain spaghetti or linguini
(see note)

½ cup halved cherry tomatoes

½ cup finely chopped red bell peppers

½ cup finely chopped summer squash

½ cup sliced pitted ripe olives

½ cup finely chopped cucumber

½ cup finely chopped ham (optional)

2 tablespoons olive oil

6 tablespoons apple cider vinegar

1. Prepare the pasta according to package directions.

2. In a large bowl, combine the pasta, tomatoes, peppers, squash, olives, cucumber, ham (if using), oil, and vinegar. Serve warm or cold.

Note: Any pasta shape can be used, but chickens seem to prefer the longer noodles.

Egg Salad for Chickens

6 eggs, hard-cooked (page 190), peeled, and finely chopped

¼ cup mayonnaise

1 cup wild bird seed mix

In a large bowl, combine the eggs, mayonnaise, and bird seed. Serve immediately.

Weekend Suet Cake for Chickens

1½ cups cornmeal

1 cup whole wheat flour

1 cup crunchy or smooth peanut butter

½ cup shortening

2 cups sunflower seeds, pumpkin seeds, millet, wild bird seed mix, cracked corn, peanuts, safflower seeds, dried mealworms, flaxseeds, and/or quinoa

In a medium bowl, combine the cornmeal, flour, peanut butter, shortening, and seeds, grains, and/or mealworms. Form into desired shapes on waxed paper by hand, or press into a waxed paper–lined pan and cut into shapes. Store at room temperature.

Crustless Pumpkin Pie for Chickens

1 can (15 ounces) pumpkin puree

1 cup almond milk or rice milk

2 eggs

1 teaspoon pumpkin pie spice (optional)

Preheat the oven to 325°F. Coat a 9" pie plate with cooking spray or shortening. In a large bowl, combine the pumpkin, milk, eggs, and pumpkin pie spice (if using). Pour into the pie plate and bake for 50 minutes, or until a knife inserted near the center comes out clean.

Acknowledgments

This book represents years of personal experience keeping chickens, but it also represents the accumulation of many debts of gratitude to people who have helped me along the way, especially my husband, Alan. When I was working my regular job, dealing with an unforeseen personal situation, and trying to get the book written on deadline—all three at once, not to mention all the mundane, daily slings and arrows that make up everyday life—he was ready to do anything to support me. He was patient with my early mornings and my evening exhaustion. He cooked, he cleaned, he gathered eggs, he fed and watered the chickens, he changed cat litter boxes, he reviewed legal documents, he went grocery shopping, he ran errands, he mailed contracts and photos, he kept me on track, he acted as a sounding board . . . and he loved me unconditionally during the days of stress as well as the days of joy. He told me he was proud of me. No one could ask for a better husband or partner.

I'd also like to acknowledge the advice and support of farmer and friend David Gundlach, who, with decades of experience keeping small flocks of both goats and chickens, kindly shared with me what he did to follow his calling to tend to the animals in his care in the most humane, loving ways possible.

A big thank-you goes to my father-in-law, Gene, as well, who generously "helped" me build my first coop—by building it almost completely on his own and letting me take part by hammering in nails from time to time.

Finally, I'd like to thank the amazing editorial team at Rodale for believing in this book and working so hard to bring out the best in it. I'm incredibly grateful to have been given the opportunity to work with such bright, optimistic, and supportive people.

—Lissa Lucas

First, I want to thank Henry Scanlon and Judy Curiale for taking a chance on me in 2001. You took me under your wings, you taught me about business, about the world, and about myself. You gave me the confidence to start down this path.

Sweet Lucie, my mother-in-law, you showed me backyard chickens were possible. I fell in love with your Easter Eggers, your little Sebright, and your huge Blue Cochin on sight. Thank you so much for inspiring this journey.

The early editors of my online guide to chicken care, Marlene Sarnat, Laurie Torres, Mark Scanlon, and Lynne Scanlon, I appreciate so much your time and thoughtful comments. Who could have known how many hundreds of thousands of people would eventually rely upon this guide to start their chicken-keeping adventures?

Anne Hays, having you in our corner made all the difference.

Lissa Lucas, you put forth the lion's share of the effort for this book. Your unique voice so beautifully comes through on these pages. I hope you are as proud of your efforts as I am.

Every single staff member at MPC—you ambassadors of chicken quan—you make this thing possible. You are appreciated far more than you know.

Last but not least, I'm grateful for the patience, skill, and professionalism of Karen Bolesta and the entire team at Rodale. Thank you so much for believing in us.

—*Traci Torres*

About the Authors

LISSA LUCAS

Backyard chicken expert Lissa Lucas is the head writer and marketing communications specialist for My Pet Chicken. She graduated Phi Beta Kappa and Summa Cum Laude from Marietta College, Ohio, and has worked variously as a journalist, an English teacher at a private school in China, and the communications director for a small, private university, among other positions. However, teaching people about chickens—and being perhaps the world's only chicken advice columnist—is her favorite job by far. She blogs from a tumbledown cottage on a rural ridge top in beautiful Ritchie County, West Virginia. In addition to keeping chickens, Lissa enjoys home brewing, nature photography, cooking, yoga, crocheting, writing, and gardening.

TRACI TORRES

Traci Torres is the CEO and cofounder of My Pet Chicken. She has been passionate about chickens ever since "Goldie" peeped out of her shell during a 3rd-grade classroom hatching experiment. She still considers herself a "wannabe" farmer and aims to one day steward a veritable Noah's ark of creatures. For now, she contents herself with chickens of all varieties, sizes, and provenances. She lives in rural Connecticut, along with her husband and two children.

About My Pet Chicken

The idea for My Pet Chicken was hatched in 2004, after husband-and-wife team Traci Torres and Derek Sasaki found disappointingly little how-to information on the web for keeping these brilliant and easy-to-care-for pets. Incorporated in 2005, My Pet Chicken has become the go-to resource for aspiring backyard chicken keepers. It is powered by a team comprised almost exclusively of enthusiastic chickenistas, working from their home offices around the country. Its web site, www.mypetchicken .com, receives 30 million page views annually and features resources like the "Pick a Chicken" breed selection tool, free chicken care and incubation guides, stunning photography, and a comprehensive and searchable "chicken help" database. The web site offers everything from chicks to coops, from hatching eggs to feeders, from organic chicken treats to chicken-themed gifts, and more. My Pet Chicken made the *Inc.* 5000 list of fastest-growing companies in 2012, appeared on the *Martha Stewart Show* and CNN, was profiled that same year on Bloomberg TV's *Enterprise*, was featured on the cover of *Entrepreneur Startups* magazine in spring 2013, and has been mentioned in publications like the *Wall Street Journal, Time,* the *New Yorker, Chicago Tribune,* and *Los Angeles Times,* among others.

Index

Underscored page references indicate boxed text and tables. **Boldface** references indicate photographs.

A

Add-on runs, 70
Adult chickens
   chores associated with, 138–40
   cold-weather, 145–47
   feeding guidelines for, 140–41
   grooming, 149–53
   hot-weather, 147–48
   nutrition for, 141–45
Aggression. *See also* Picking on other chickens
   from confinement, 7, 8, 60, 62
   toward newcomer, 178
   of roosters, 51
   as sign of stress, 160
AI viruses, 173
Alarm calls, about predators, 48, 64, 92, 96, 99
American Poultry Association (APA), 6
Amprolium, in starter feed, 120
Anconas, laying rate of, 30
Anthelmintics, for worm treatment, 170
Appenzeller Spitzhauben, **16**
Araucana, **16**
Auctions, buying from, 43
Automatic coop doors, 56, 67–68, 97
Automatic feeders and waterers, 68
Avian influenza (AI), 173
Avocados, 227

B

Baby chicks, **41, 45, 47, 81, 103, 106, 108, 110, 113, 128**
   arrival of, 123–24, 126–28
   broody hens raising, 77, 78–79, 134
   children and, 74–75, 77, 132
   grit for, 120–21
   hand-taming, 134–35, **134**
   heat sources for, 117–19, **118**, 132–33
   injured, separation for, 124–25
   introducing
      to flock, 179
      to outdoors, 133–34
   loss of (*see* Losses)
   ordering (*see* Ordering baby chicks)
   pasting on, 127, 128, **128**, 130, 131
   quarantine unnecessary for, 167
   reserving (*see* Reservations for baby chicks)
   sexed, 44, 45
   sources of, 45–47, **121**
   space requirements for, 133
   starter feed for, 119–20, 135, 141, 142
   starting with
      ease of, 43–45
      reasons for, 40
   transitioning, to coop, 113, **113**, 114, 135–37
   umbilical cord on, 128
   vent-sexing errors with, 41
   water for, 121, 128
   weak, managing, 129–32
Backyard chicken keeping
   advantages of, 8, 9
   avoiding problems with, 11–13
   reasons for choosing, 10–11, 10, 172
Bacterial contamination, of eggs, 146, 185, 186
Baking, with backyard eggs, 187, 188
Bantams
   eggs of, 131, 188
   feed and water access for, 179
   as kid-friendly, **112**
   mash feed for, 119
   in mixed flocks, 39
   Sebright, **16**
   sexed, 44
   Silkie, **16, 112**
   story about, 130–31
Basic recipes
   Best-Ever Scrambled Eggs, 190
   Deviled Eggs, Five Ways, 191
   Dressed-Up Egg Salad, 193

Basic recipes *(cont.)*
    Foolproof Hard-Cooked Eggs, 190
    Homemade Sesame Mayonnaise, 193
    Red Pickled Eggs, 192, **192**
Bathing chickens, 169
Battery hens, rescuing, 43
Beak trimming, 7, 43, 151–53
Bedding
    for adult chickens, 139
    for baby chicks, 122–23, **123**, 129
    saving money on, 122
Behavior problems. *See* Misbehavior
Behaviors, normal, mistaken for misbehavior,
        179–81
Benign symptoms, observed by new chicken
        keepers, 161–65
Biosecurity precautions, 165–67
Bird flu, 173–74
Birds of prey, as predators, **86**, 90–91, **91**, 92, 98
Blind chickens, light perceived by, 46, 72
Blood spots, on or in eggs, 162, 166
Bloom, on eggs, 185, 186
Blue Ameraucanas, **41**
Blue eggs, 11, 32, 33, 33
Boredom
    from confinement, 60, 61, 62, 62, 133
    preventing, in cold-weather chickens, 147
Brahma, **16**
Breakage, of beak, 153
Breakfast recipes
    Cream Cheese Eggs in Crispy Ham Crust, 194,
        194
    Dropped Blueberry Scones with Lemon Curd, 201
    Dutch Baby Puffed Pancake, 200
    Eggs Florentine with Pesto, 198
    Eggs in Rings, 195
    Green Breakfast Wraps, 200
    Ham Strata, 199
    Italian-Style Eggs, 195
    Lissa's Huevos Rancheros, 196, **196**
    Lissa's Sunny Egg Clouds, 198
    Pineapple-Stuffed French Toast, 199
    Poached Eggs on Tomato-Eggplant Beds, 197
Breast lump, on chicken, 162
Breeders
    buying baby chicks from, 47
    as NPIP participants, 48
Breeding standards, 4, 6

Breeds
    broody, 73, 79
    docile, for children, **112**
    economical eaters among, 141
    escaping predator attacks, 96
    fancy, adding to homogeneous flock, 179
    friendliness of, 134
    in "hatchery choice" assortments, 103
    hatching times of, 111
    limited, with started birds, 43
    mixing, 38–39, **38**
    reserving, 104, 113
    selecting (*see* Breed selection)
    standardizing, 6
    substitute, with ordering problems, 105, 111, 112
    verifying correct shipment of, 126–27
    voices of, 98
Breed selection
    appearance considerations for, 15, **16–17**
    general considerations for, 14–15, 18–27
        broodiness, 28–30, **29**
        cold hardiness, 34–35, **34**
        docility, 36–37, 36
        egg color, 32–34, **32**
        egg-laying rate, 30–31, **31**
        egg size, 31–32
        fancy feathering, 37–38
        foraging, 35
        heat hardiness, 35
        winter laying, 35
Breed Selector Guide, 14–15, 18–27
Brinsea EcoGlow brooder, 75–76, 77, 117, 137
Brooder (hen), 73. *See also* Broodiness; Broody hens
Brooder (housing)
    basic requirements for, 74–76, 126
    bedding for, 122–23, **123**, 129
    cats and, 74–75, 77, 82
    checking on, 128–29
    in coop, 71
    creating, 73
    fire hazard in, 75, 78
    heating, 12–13, 75–76, 77, 78, 117–19, **118**, 132–33
    losses occurring in, 108
    purpose of, 72–73
    siting, 76–77
    space recommendations for, 75, 133
    when to order supplies for, 104
Brooder kits, 73

Broodiness, **29**. *See also* Broody hens
  encouraging, 184
  meaning of, 28–30, 73
  signs of, 159, <u>160</u>, 163
  of specific breeds, <u>19</u>, <u>21</u>, <u>23</u>, <u>25</u>, <u>27</u>
Broody hens. *See also* Broodiness
  bothered by egg gathering, <u>83</u>, 184
  caring for baby chicks, 77, <u>78–79</u>, 134
  as economical eaters, 141
  introductions eased by, 115
  mite infestations in, 159
  modern poultry keeping and, 6
  older birds as, 115–16
Brown eggs, <u>11</u>, 32, 33, <u>33</u>, 34, 164
Bucket nests, 69
Built-in nests, 69, **70**
Burials, 128, 174–75

C

Calcium
  deficiency of, <u>160</u>
  in feed, 119, <u>142</u>, 144, 145
  in oyster shell, 142
  pimples, on eggshell, <u>148</u>, 161
Camouflage, egg color providing, 32
Cancellation policies, of hatcheries, 103–4
Candling, 33, 162, <u>166</u>
Cats
  adult chickens and, 82
  protecting brooder from, 74–75, 77, 82
Cautions, about specific breeds, <u>19</u>, <u>21</u>, <u>23</u>, <u>25</u>, <u>27</u>
Cedar shavings, avoiding, for bedding, 122
Chalazae, <u>150</u>
Change policies, of hatcheries, 103–4
Checklist, chicken readiness, <u>xix</u>
Chicken readiness checklist, <u>xix</u>
Chicken tractors. *See* Tractor-style coops
Chick feed, 119–20, 135
Chick waterer, 76
Children
  chicken keeping benefiting, 11
  docile breeds for, **112**
  pet losses and, <u>172</u>, 174, **174**
  protecting baby chicks from, 74–75, 77, 132
Cleaning
  chicken-keeping equipment, 165

  of coop, 139–40, 147, 165, 171
    flock management types and, <u>55</u>, 60, 62
Coccidiosis, 120, 128, 129
Cochins, 4, <u>79</u>, **112**, **115**
Cold hardiness, 34–35, **34**
  of specific breeds, <u>18</u>, <u>20</u>, <u>22</u>, <u>24</u>, <u>26</u>
Cold weather
  preparing chickens for, 145–47
  reduced laying during, 71
  transitioning to coop during, 136–37
Combs
  cold-weather protection for, 147
  spots on, 163
Commercial chicken feed, 6, 7, 141, 142, 144
Communication
  of baby chicks, 118, 127
  between chickens, 47, 48–49, <u>96</u>, <u>109</u>, 116
Compost material, 63, 129
Confined ranging, pros and cons of, 56–58
Confinement, problems of, 7
Convulsing chickens, 163
Cooking with backyard eggs, 187–88
Coop doors
  automatic, 56, 67–68, 97
  opening and closing, 54–56, <u>58</u>, 59, 63, 138
Coops
  bedding for, 122–23, 139
  chores associated with, 138–40, 147
  cleaning, 139–40, 147, 165, 171
  cold-weather chickens and, 145–46, 147
  desirable features for, 67–71
  fire hazards in, 71, 75, 113, 136
  heating, 113, 136, 137, 145–46
  hospital area in, 71, 167
  light in, 70, 71–72, **72**
  predator-proofing, 95–96
  separating newcomers in, 177–78
  siting, 66–67
  time needed to prepare, 112
  transitioning baby chicks to, 113, **113**, 114, 135–37
  ventilation in, **68**, 70, 95, 145, 148
  when to buy, 13
Corvids, as predators, 91–92
Cougars, as predators, 93–94
Coyotes, as predators, 92–93, **93**
Crate nests, 69
Cremation, pet, 175
Crested birds, dangers faced by, 37–38, 39, 96, 151

Crest trimming, 151
Cross beak, 152, <u>152</u>
Crows, as predators, 91–92
Crumbles, 119, 141
Crying
    by baby chicks, 118, 127
    as sign of stress, <u>160</u>
Cuckoo Marans, docility of, 36

D

Dairy products, 227
Dancing, by roosters, 181
D'Anvers Hen, **17**
Deaths. *See* Losses
Deep-litter method, for reducing coop cleaning, 139, 140
Delayed chick orders, 105, 110–11
Dessert recipes
    Angel Food Cake, 224
    Baked Barley-Cherry Custard, 221
    Berries and Cream Sponge Cake, 226
    Bread and Butter Pudding, 223
    Classic Chocolate Soufflé, 221
    Crème Brûlée, 222
    Epiphany Lemon Bars, 219
    Honey-Vanilla Ice Cream, 224
    Lemon Poppy Seed Pound Cake, 225, **225**
    Lissa's Solstice Egg Nog, 219
    Profiteroles, 220
    Pumpkin Custard, 222
Diatomaceous earth, for infestations, <u>168</u>
Digestion, grit for, 121, 135, 142
Docility, 36–37
    of specific breeds, <u>19</u>, <u>21</u>, <u>23</u>, <u>25</u>, <u>27</u>, <u>36</u>
Dogs
    as predators, 74, 81–82, <u>84</u>
    training, to accept chickens, <u>84–85</u>
Domestication of chickens, 1, <u>92</u>
Draft protection, for brooder, 75
Dust, from indoor brooders, 77
Dust bathing, 163

E

Easter Eggers, **17**
    cross-beaked, <u>152</u>
    docility of, 36, **47**

as poor winter layers, 34
    roosters and, 51
    story about, <u>130–31</u>
    traits of, <u>146</u>
EcoGlow brooder, 75–76, 77, 117, 137
Egg appearance, benign issues with, 161–62, 163–64
Egg color, <u>11</u>, 32–34, **32**
    from bantams, <u>131</u>
    earlobe color and, <u>127</u>
    lightening of, 162
    of specific breeds, <u>18</u>, <u>20</u>, <u>22</u>, <u>24</u>, <u>26</u>, <u>146</u>
Egg eating, as sign of stress, <u>74</u>, <u>160</u>
Egg gathering
    chickens unbothered by, <u>83</u>
    fun of, 138
    during owner's absence, <u>74</u>
    tips on, 184
    winter, 146
Egg hatching
    best temperature for, <u>29</u>
    broodiness for (*see* Broodiness; Broody hens)
    by broody hens, <u>78</u>, <u>79</u>
    delayed, 105
    at home
        difficulties with, 40–41, <u>42</u>
        roosters needed for, 50
    incubators for, 6, 28, 40, <u>42</u>, 105, 111, 125
    time needed for, <u>111</u>
Egg irregularities, <u>148</u>
Egg laying
    average age at, <u>44</u>
    decreased with age, 114, <u>173</u>
    light triggering, <u>46</u>
    molting slowing, 31, 34, 35
    rate of, 6, 30–31, **31**
        in specific breeds, <u>18</u>, <u>20</u>, <u>22</u>, <u>24</u>, <u>26</u>
    signs of readiness for, <u>67</u>
    stopped by broody hens, 29, 30
    winter (*see* Winter layers; Winter laying)
Egg quality, factors affecting, 8, 9, 11, 32
    flock-management style, <u>55</u>, 56, 57, <u>60</u>, 62
Egg quantity comparison chart, <u>188</u>
Egg recipes. *See also* Basic recipes; Breakfast recipes; Dessert recipes; Main dish recipes; Side dish recipes
    egg quantities for, 188, <u>188</u>
Eggs, symbolism of, 4

Egg size, 31–32, <u>148</u>, 161
    classifications of, 187
    of specific breeds, <u>18</u>, <u>20</u>, <u>22</u>, <u>24</u>, <u>26</u>
Egg skelter, for storage, 187
Egg washing, pros and cons of, 185–86
Eglu Cube tractor, 57
Electricity, for coops, 71–72, **72**
Electrolyte formulas, 121
Equipment, cleaning and disinfecting, 165
Evans, Chris, 48
Exhibition chickens, 4, 14, 38, 51
Extra chickens
    sent with orders, 106–7
    when to order, 109–10

F

Facial recognition, by chickens, <u>133</u>
Factory farm "free range," <u>61</u>
Factory farms
    age of slaughter in, 115
    conditions of, 7–8
    egg washing by, 185
    medicated feed in, 120
    reducing dependence on, 114
    salmonella outbreaks in, 172
Fairy eggs, 161, **186**
Fall chicks, 113–14
Family size, number of hens needed for, 31
Fancy feathering, drawbacks of, 37–38
Faverolles
    in mixed flocks, 39, 179
    voices of, 48, <u>98</u>
Feather clipping, 150–51, **151**
Feather loss, causes of, <u>168</u>, 169–71, <u>170</u>. *See also*
        Molting
Feathers
    fancy, drawbacks of, 37–38
    left after predator attacks, 89, **89**
Feed
    for adult chickens, 119, 141
    for baby chicks, 119–20, 135, 141
    commercial, 6, 7, 141, 142, 144
    for different-age birds, 143–44
    estimating amount needed, 140–41
    homemade, 144–45
    hormones disallowed in, 141

    predator-proofing, 99
    for stress relief, 169
    supplying, during owner's absence, <u>74</u>
    when to swap, <u>142</u>
Feed costs
    estimating, 140
    with free ranging, <u>55</u>, 62
    with part-time ranging, <u>60</u>, 62
    with permanently sited coops, <u>58</u>
Feeders
    additional, when expanding flock, 179
    automatic, 68
    for baby chicks, 76
    for cross-beaked chickens, 152, <u>152</u>
    dirtying of, 128–29
    predator-proofing, 99
    refilling, for adult chickens, 138–39
    wasted feed and, 140
Feed stores, buying baby chicks from, 46
Fenced yard, for confined ranging, 56
Fertile eggs
    risks of buying, 40–41, <u>42</u>
    roosters needed for, 50
    when to buy, 40
Fertilized eggs, as edible, <u>37</u>
Fertilizer. *See* Manure, chicken
Fighting, 180–81
Fire hazards, 71, 75, <u>78</u>, 113, 136
Fishy-smelling eggs, 164
Flighty, meaning of, 36
Flock-management styles
    confined ranging, 56–58
    free ranging, 54–56, <u>55</u>, **57**, **63**, <u>64–65</u>
    full-time confinement, 61, <u>62</u>
    guidelines for choosing, 62–63
    part-time ranging, 59–61, <u>60</u>
Flocks
    dynamic and culture of, 47–50, 116
    establishing (*see* Ordering baby chicks)
    introducing new birds to, 49–50, 51, 115
        guidelines for, 177–78
        mistakes to avoid when, 178–79
        quarantine for, 165–67, 178
    methods of starting
        baby chicks, 43–47
        home hatching, 40–41, <u>42</u>
        juvenile chickens, 41, 43
        rescued battery hens, 43

Flocks *(cont.)*
  planning considerations for, 47
  protecting, from predators, 95–99 (*see also*
      Predators)
  roosters in, 50–51
Flyers, 36–37
Food choices, controlling personal, 10
Foragers, 35
  specific breeds as, <u>19</u>, <u>21</u>, <u>23</u>, <u>25</u>, <u>27</u>
Foraging, 56, 135
Foxes, as predators, <u>64–65</u>, 88–89
Free choice feeding, 143, 169
Free ranging
  factory farm–style, <u>61</u>
  free choice feeding with, 143
  grit from, 142
  predator threats to, <u>64–65</u>
  pros and cons of, 54–56, <u>55</u>, **57**, **63**
  for stress relief, 169
Full-time confinement, pros and cons of, 61, <u>62</u>
Funeral, for pet chicken, 174–75

G

Garlic, 227
Gathering eggs. *See* Egg gathering
Genetic benefits, from older hens, 116
Gizzard, 121, 142, 162
Gray eggs, 32
Green eggs, <u>11</u>, 33
Green roof, on coops, 68
Grit, 120–21, 135, 137, 142, 162
Grooming concerns
  bathing, 149
  beak trimming, 151–53
  blunting spurs, 153
  crest trimming, 151
  nail trimming, 149–50
  wing clipping, 150–51
Grower/developer feed, 141, 144

H

Handling chickens, <u>126</u>
Hand-taming baby chicks, 134–35, **134**
Hand washing, 132, 165, 171

Hardware cloth, for securing run, 96, 97
Hatcheries
  buying from, 45–46, 165
  NPIP, <u>48</u>, 165, 167
  ordering from (*see* Ordering baby chicks)
"Hatchery choice" assortments, problems with, 103
Hatching eggs. *See* Egg hatching
Hay bedding, avoiding, 122
Head-pecking, 180, 181
Heat
  for brooders, 12–13, 75–76, 77, <u>78</u>, 117–19,
      118, 137
  for coops, 71, 113, 136, 137, 145–46
  for weak chicks, 130
Heat hardiness, 34, 35
  of specific breeds, <u>19</u>, <u>21</u>, <u>23</u>, <u>25</u>, <u>27</u>
Heat stress, egg lightening from, 34
Hens
  acquiring male characteristics, <u>63</u>, <u>73</u>
  adding, to flock, 178
  battery, rescuing, 43
  broody (*see* Broody hens)
  good-laying, signs of, <u>41</u>
  juvenile, appearance of, <u>7</u>
  normal behaviors of, 180, 181
  number of, per rooster, 50, 170, 180
  older, benefits of keeping, 114–16
  roosters mating with, <u>105</u>, 170–71, 180, 181
  symbolism of, 3–4
  teaching babies about food, <u>75</u>
Hen saddle, 170
History of chickens, 1–4, 6–8, <u>88</u>, <u>92</u>
Hobby, American chickens as, 2, 4, 6
Home exam, for detecting illness, 158
Hormones, disallowed in feed, 141
Hospital area, in coops, 71, 167
Hot-weather chickens, encouraging, 147–48
Housing. *See* Brooders; Coops
Hunters, chickens as, 35

I

Illnesses
  biosecurity preventing, 165–67
  chickens hiding, 156, 166, **167**
  in confined flocks, 61
  egg lightening from, 34

home exam detecting, 158
information for vet about, 159, 161
in newcomer birds, 166–67
quarantine with, **159**, 160
seeking treatment for, 132
separation for, 125
signs of
    checking for, 129, 173
    common, 157, 157, 158
signs unrelated to, 157, 157, 159
stress-related, 160
when to seek help for, 158–59
Impaction of crop, 142–43, 158, 162
Incubators, 6, 28, 40, 42, 105, 111, 125
Infestations
    in bedding materials, 122
    biosecurity preventing, 165–67
    in confined flocks, 61
    intestinal parasite, 170
    lice, 167, 168
    mite, 153, 159, 167, 168, 169
    scaly leg mites, 158, 167, 169
Injuries
    broken beak, 153
    from fighting, 181
    hospital area for, 71
    protecting chicks with, 124–25
Insect control, chickens providing, 11, 35, 80
Insulation, coop, 145
Intestinal parasites, 170

J

Juvenile chickens
    feather clipping and, 150
    quarantine unnecessary for, 167
    starting with, 41, 43

L

Langshan, **17**
Latches, for protecting coop, 95
Layer feed, 140, 141, 142, 143, 144, 170
Laying cycles, light triggering, 46
Laying eggs. *See* Egg laying
Leg feathers, weather affecting, 38
Leg mites, scaly, 158, 167, 169

Lice infestations, 167, 168
Life span of chickens, 114
Lift-up roof, on coops, 71
Light
    in coops, 70, 71–72, **72**
    laying cycles and, 46
Listlessness, as sign of stress, 160
Live-arrival guarantee, 124–25
Liver problems, 150, 153
Losses
    children coping with, 172, 174, **174**
    guarantees against, 124–25
    managing, 109–10
    notifying hatchery about, 128
    ordering policies about, 104
    pecking order changed by, 115
    reasons for, 125–26
    risks of, 105–6, 107–8, 125
Lump, on chicken breast, 162
Lynx, as predators, 93–94

M

Main dish recipes
    Croque Madame, 209
    Crustless Leek and Goat Cheese Quiche, 208
    Everyday Soufflé, 207
    Extra-Cheese Monte Cristo, 208
    Lissa's Versatile Quiche, 206
    Provençal Mushroom-and-Vegetable Main Dish
        Salad, 211
    Quick Baked Scotch Eggs, 202
    Rumbledythumps, 204
    Salmon Patties with Cucumber Ribbons and
        Dynamite Sauce, 210, **210**
    Scotch Borders Golden Egg Nests, 204
    Scotch Glazed Meatloaves, 203, **203**
    Sesame Ginger Tuna Salad, 205
    Smoked Salmon, Asparagus, and Goat Cheese
        Scrambler, 205
    West Virginia Quiche-Stuffed Pattypan Squash,
        202
Makeshift nests, 69
Malnourishment, 143
Manure, chicken, 10, 61, 62, 129, 132, 139
Marek's disease, 106
Mash, 119, 140, 141

Mating, 105, 170–71, 180, 181

Mealworms, as treats, 135

Medicated feed, 120, 144

Medication, in drinking water, 121

Misbehavior. *See also* Aggression

    normal behaviors mistaken for, 179–81

    during owner's absence, 74

Mite infestations

    beak weakness from, 153

    in broody hens, 159

    managing, 168

    in newcomer birds, 167

    scaly leg mites, 167, 169

Mixing breeds, 38–39, **38**

Modern Game, **17**

Modern poultry keeping, 6–8

Molting

    feather loss with, 162, 169

    laying slowed during, 31, 34, 35

    light affecting, 72

    protein for, 143

    water needs and, 148

    wing clipping after, 150

My Pet Chicken

    brooder kit of, 73

    business of, viii, 11

    care information from, 101

    chick feed of, 119

    customers of, 101

    juvenile breeds from, 43

    live-arrival guarantee of, 124

    loss policy of, 128

    as NPIP participant, 48

    ordering policies of, 103, 106, 107, 111, 126

    sexed bantams from, 44

    sexing-accuracy guarantee of, 109

    shipping practices of, 75, 114

N

Nail trimming, 149–50

National Poultry Improvement Plan (NPIP), 43, 45, 47, 48, 165, 167

Nest boxes, 68–69, **68**, 170, 179

Newcomer birds

    introducing, to flock, 49–50, 51, 115

    guidelines for, 177–78

    mistakes to avoid when, 178–79

    quarantine for, 165–67, 178

Newspaper bedding, avoiding, 123

NPIP. *See* National Poultry Improvement Plan

Nutritional deficiencies, 160, 170

Nutritional needs, 141–45

O

Older hens, benefits of keeping, 114–16

Omnivores, chickens as, 35, 139

Onions, 227

Opossums, as predators, 83, 86

Order changes, problems with, 103

Ordering baby chicks

    considerations before

        handling delays, 110–11

        handling substitutions, 111–12

        ordering extras, 107–10

        time of year, 112–14

    mistakes to avoid when, 102–7

Orpingtons, **16**, 36, 51, 79

Outside egg door, on coops, 69–70

Oyster shell supplement, 141, 142, 144

P

Paper towel bedding, avoiding, 123

Parasites. *See also* Infestations

    intestinal, 170

Part-time ranging, pros and cons of, 59–61, 60

Pasting, on baby chicks, 127, 128, **128**, 130, 131

Pecking, feather loss from, 169

Pecking order

    behavior in, 160, 180

    birds low in, 7, 38, 39

    changes disrupting, 49–50, 177

    crest trimming and, 151

    illness affecting place in, 156

    maintaining stability in, 115, 116

    meaning of, 47–48, 176–77

    of roosters, 51

Pelletized feed, 110, 140, 141

Permanently sited coops, 58, 59, 66–67

Personality of chickens, 10, 15, 36–37, 49, 49, 134

Pest control, chickens providing, 11, 35, 80

Pesticides, for infestations, 168
Petroleum jelly
    for protecting combs, 147
    for treating scaly leg mites, 169
Pets, chickens as, 10, 11, 14, 56
Picking on other chickens. *See also* Aggression
    space preventing, 74, 151
    splayed leg deformity and, 123
    stress causing, 160, 169
Pineal gland, for light detection, 72
Pine wood shavings, for bedding, 122, 139
Pink eggs, 11
Plastic nest boxes, 69
Poisoning predators, avoiding, 99
Polish hens, **16**, **17**
    large crests of, 151
    laying rate of, 30
    in mixed flocks, 39, 179
Poo, dried or sticky, 127, **128**, 164
Post office, chick deliveries from, 123–24
Potato skins, green, 227
Predators
    alarm calls about, 48, 64, 92, 96, 99
    crested birds and, 38, 151
    full-time confinement preventing, 61
    myth about, 80
    protection from
        in coop, 69–70, 95–96
        general measures for, 98–99
        by opening and closing coop door, 54–56, 58,
            59, 63, 97
        during owner's absence, 74
        with part-time ranging, 60
        in run, 96–98
    types of
        birds of prey, **86**, 90–91, **91**, 92, 98
        cougars, 93–94
        coyotes, 92–93, **93**
        crows and other corvids, 91–92
        domestic dogs, 81–82, 84
        foxes, 64–65, 88–89
        house cats, 82
        lynx, 93–94
        opossums, 83, 86
        raccoons, 70, 82–83, 87, 95
        rats, 88
        snakes, 86–88
        snapping turtle, 94–95

weasel family, 89–90
wolves, 92–93
weaknesses attracting, 156
Pregnancy, chicken keeping during, 171
Protein
    deficiency of, 160, 170
    layer feed lacking, 119
    needs for, 143, 145, 152
Pullets
    average egg-laying age of, 44
    layer feed for, 142

Q

Quarantine
    for new birds, 165–67, 178
    for sick birds, **159**, 160

R

Rabies, from raccoons, 83
Raccoons, as predators, 70, 82–83, 87, 95
Rats, as predators, 88
Recipes. *See also* Basic recipes;
        Breakfast recipes; Dessert recipes;
        Main dish recipes; Side dish recipes;
        Treats for chickens, recipes
    egg quantities for, 188, 188
    for homemade chicken feed, 144
Refrigeration, of eggs, 186–87
Regulations
    on chicken keeping, checking, 13, 102–3
    on pet burials, 174–75
    on washing and storing eggs, 185–86
Reputation of chickens, 2
Rescue birds
    battery hens as, 43
    quarantining, 165, 166–67
Reservations for baby chicks, 12, 13
    checking orders after, 126, 127
    problems with canceling, 13, 103
    scheduling, 113, 114, 123
    of trendier breeds, 104
Rhode Island Reds, 179
    aggressive roosters among, 51
    laying rate of, 30

Rhode Island Reds (cont.)
    picking on other hens, 37
    voices of, 48–49, 98
Rodents
    chickens catching, 80
    feed lost to, 141
    protection from, 74, 77, 99
Roll-away nest boxes, 69
Rollin, Bernard, 49
Roofs
    on coops, 68, 71, 148
    for runs, 60
Rooster of Basel case, 94
Roosters
    average crowing age of, 116
    favorite breeds of, 51
    feather loss caused by, 168, 170–71
    juvenile, appearance of, 7
    mating with hens, 105, 170–71, 180, 181
    multiple, rules for keeping, 50–51
    normal behaviors of, 180–81
    number of hens for, 50, 170, 180
    pros and cons of, 50
    spurs of, **51**
        blunting, 153
    symbolism of, 3
    treating infestations in, 168
    warning of predator attacks, 64, 92, 99
Roosts, for coops, 68, **68**
Roundworms, 170
Runs
    add-on, 70
    for confined ranging, 56
    for part-time ranging, 59, 60
    predator-proofing, 96–98
    roof for, 60
    in tractor-style coops, 57

S

Salmonella, 171–73
Salt, limited, for chickens, 227
Sanitation
    of coop, 139–40
    for salmonella protection, 171
Scaly leg mites, 167, 169
Scissor beak, 152, 152
Sebright Bantam, **16**

Separation, for injured or sick chicks, 124–25
Seramas, **17**, 187
Sex determination of chickens, 81
Sexing errors, in orders, 104, 105, 108–9
Shade, for hot-weather chickens, 148
Shamo, **16**
Shampoos, 169
Shipment of chicks, 106, 107
    expecting arrival of, 123–24
    losses during, 108
Shipped chicks, broody hens raising, 79
Shoes, for working with flock, 165
Side dish recipes
    Corn Custard, 218, **218**
    Egg Foo Yung, 217
    Egg Noodles, 213
    Fried Brown Rice, 215
    Grandma Waterman's Potato Salad, 214, **214**
    Potato and Onion Frittata, 215
    Quick Egg Drop Soup, 213
    Salad Niçoise, 212
    Zucchini alla Napoli, 216
Silkie Bantam, **16**, **112**
Silkies
    as broody breed, 79
    docility of, 36
    large crests of, 151
    laying rate of, 30
    nail problems of, 149
    voices of, 49, 98
    White, **179**
Skunks, 89
Sleep, passive state during, 120
Snacks, for baby chicks, 119
Snakes, as predators, 86–88
Snapping turtles, as predators, 94–95
Space recommendations
    for adding to flock, 179
    for brooder, 75, 133
    during owner's absence, 74
    for preventing aggression, 74, 151
    for preventing stress, 165, 169
    for roosters, 50–51
Speckled eggs, 32
Splash Ameraucana, **41**
Splash Silkie chick, 81
Splayed leg, from newspaper bedding, 123
Spring chicks, 114

Spurs, rooster, **51**
blunting, 153
Star-San sanitizer, 140
Started pullets, starting with, 41, 43
Starter feed, 119–20, 135, 141, _142_
Starter/grower feed, _142_, 144
Starting a flock. _See_ Flocks, methods of starting
Sticky poo, 164
Storage, egg, 186–87
Straight-run birds, 44–45, 46, 47
Straw bedding, avoiding, 122
Stress
from confinement, 7
eggshell damage from, _180_
feather loss from, 169–71
infections and, 120
laying rate reduced by, 31
signs of, _160_
space preventing, 165, 169
Substitutions, for ordering problems, 105, 111–12
Suet cake, 147, 231
Sugar, limited, for chickens, 227
Sugar water, for weak chicks, 130–31
Sulfamethazine sodium solution (Sulmet), for
coccidiosis, 120
Sultan, **16**
Sunbathing, 163, **163**
Supplements, 141–43
Supplies, when to get, 104
Surprise gifts, chickens unsuitable for, 104–5
Sussex chickens, docility of, 36
Sustainable living, 10
Symbolism of chickens, 3–4

T

Tapeworms, _170_
Taste, of backyard eggs, 11, 183, 189
Taxoplasmosis, 171
Thermometer, for monitoring brooder
temperature, 118
Time commitment
for care of chickens, 13
for free ranging, 56
Tractor-style coops
for confined ranging, 56–58
for part-time ranging, 59

pros and cons of, _59_
siting, 66
Transportation, for shipping chickens, 6, 9
Treats for chickens, 61, 134–35, 147, 148, 173
recipes
Crustless Pumpkin Pie for Chickens, 231
Egg Salad for Chickens, 230
Hatchday Cake for Chickens, 228
ingredients to exclude from, 227
Pasta Salad for Chickens, 230
Porridge for Chickens, 229
Scrambled Eggs for Chickens, 228
Summer Fruit Salad for Chickens, 229
Waldorf Salad for Chickens, 229
Weekend Suet Cake for Chickens, 231
Tyrannosaurus rex, chickens related to, _90_

U

Umbilical cord, on baby chicks, 128

V

Vaccinations, 46, 47, 106, 120
Ventilation, coop, **68**, 70, 95, 145, 148
Vents, pasting of, 127
Vent sexing, 41, 43, 108–9
Voices/vocalizations, 48–49, _96_, _98_

W

Walk-in coop, 70–71
Washing eggs, pros and cons of, 185–86
Washing hands, 132, 165, 171
Washing hands, after handling chicks, 132
Water
for baby chicks, 121, 128
for hot-weather chickens, 148
during owner's absence, _74_
preventing winter freezing of, 146–47
Waterers
additional, when expanding flock, 179
automatic, 68
chick, 76
for cross-beaked chickens, 152

Waterers *(cont.)*
  dirtying of, 128–29
  heated, 67, 146
  predator-proofing, 99
  refilling, 138–39
Weasel family predators, 89–90
Welsummer breed, 141
White eggs, 32
White-Faced Black Spanish, **17**
White Plymouth Rock, **34**
White roof, on coop, 148
White Silkie, **179**
Windows, in coop, 70, 95
Wing clipping, 150–51, **151**
Winter layers, 35
  poor, 30, 34, 141
  specific breeds as, <u>18</u>, <u>20</u>, <u>22</u>, <u>24</u>, <u>26</u>

Winter laying
  light and, 71–72
  sporadic, 114
Wolves, as predators, 92–93
Wooden nests, 69
Worm infestations, <u>170</u>

Y

Yogurt, 131, 152, 227
Yokohama, **17**